The Scientific Method

The scientific method delivers prosperity, yet scientific practice has become subject to corrupting influences from within and without the scientific community. This essential reference is intended to help remedy those threats. The authors identify eight essential criteria for the practice of science and provide checklists to help avoid costly failures in scientific practice. Not only for scientists, this book is for all stakeholders of the broad enterprise of science. Science administrators, research funders, journal editors, and policymakers alike will find practical guidance on how they can encourage scientific research that produces useful discoveries. Journalists, commentators, and lawyers can turn to this text for help with assessing the validity and usefulness of scientific claims. The book provides practical guidance and makes important recommendations for reforms in science policy and science administration. The message of the book is complemented by Nobel Laureate Vernon L. Smith's foreword and an afterword by Terence Kealey.

J. Scott Armstrong has dedicated his life to discovering useful scientific findings. During his 52 years as a professor at the Wharton School, University of Pennsylvania, he has delivered over 100 lectures at universities in 27 countries, won numerous awards, and published four books and over 400 management science articles.

Kesten C. Green has published nearly 40 scientific works, mostly concerned with improving the predictive validity of methods and models. Before becoming an academic to follow his passion for research, he was an entrepreneur for 30 years. His pursuit of useful knowledge has been a common thread through his career.

The Scientific Method

A Guide to Finding Useful Knowledge

J. Scott Armstrong
Wharton School, University of Pennsylvania

Kesten C. Green
University of South Australia

CAMBRIDGE
UNIVERSITY PRESS

University Printing House, Cambridge CB2 8BS, United Kingdom

One Liberty Plaza, 20th Floor, New York, NY 10006, USA

477 Williamstown Road, Port Melbourne, VIC 3207, Australia

314–321, 3rd Floor, Plot 3, Splendor Forum, Jasola District Centre, New Delhi – 110025, India

103 Penang Road, #05-06/07, Visioncrest Commercial, Singapore 238467

Cambridge University Press is part of the University of Cambridge.

It furthers the University's mission by disseminating knowledge in the pursuit of education, learning, and research at the highest international levels of excellence.

www.cambridge.org
Information on this title: www.cambridge.org/9781316515167
DOI: 10.1017/9781009092265

First published 2022

A catalogue record for this publication is available from the British Library.

ISBN 978-1-316-51516-7 Hardback
ISBN 978-1-009-09642-3 Paperback

Cambridge University Press has no responsibility for the persistence or accuracy of URLs for external or third-party internet websites referred to in this publication and does not guarantee that any content on such websites is, or will remain, accurate or appropriate.

Michael J. Mahoney

(1946–2006)

Michael J. Mahoney dropped out of high school and worked as a psychiatric aide until his allergist encouraged him to relocate to Arizona. He received an undergraduate degree at Arizona State University in 1969 and, later, a PhD in psychology from Stanford. He joined the faculty at Penn State University in 1972, where he remained until 1985. Following his time at Penn State, he taught at the University of California, Santa Barbara, for 5 years then, in 1990, he moved to the University of North Texas where he remained for 15 years.

Michael was a skeptic and used his skepticism to study current scientific beliefs via experimentation. He was creative in finding important problems to study. For example, when he took up weightlifting, he wrote papers on sports psychology. He also became a national champion weightlifter and served as a resident psychologist to the US weightlifting team in preparation for the 1980 Moscow Olympic Games.

Michael was, as far as we are aware, the first scientist to undertake a program of research in which scientists were the subjects. His book, *Scientist as Subject: The Psychological Imperative*, published in 1976, led the way to the scientific study of scientific practices.

Scott became interested in Michael's research early on. It inspired his first paper on scientific practice, "Advocacy and Objectivity" (Armstrong, 1979). Although they did communicate with each other by mail and email, they never met, to Scott's regret. Scott regards Michael as the person who had the greatest influence on his work on the scientific method.

Michael died on May 31, 2006 at the age of 60, in the prime of his career. He had published more than 250 scholarly articles in psychology, authored and edited 19 books, was editor of four journals, and served on 23 editorial boards.[1] He made many important scientific findings and helped others with their research.

Rest in peace, Michael. You were fearless in your search for truth.

[1] Biographical details drawn from Bandura (2008), Marquis et al. (2009), Smucker (2008), and Warren (2007).

CONTENTS

TABLES AND CHECKLISTS

Tables

Checklists

FOREWORD

by Vernon L. Smith

I found this book to be a particularly engaging and useful treatment of scientific method and practice whereby the authors' target is to improve practice. There is a lot of subversion out there that they want to avoid with a content-positive approach. The book's themes for improving scientific practice are generated from earthy checklists, all concerned to improve compliance with the substantive content of science. It results in checklists for eleven user categories, from researchers to courts.

The checklists all derive from key elements of the scientific method, summarized by eight criteria. I want to apply seven of these elements to my early work in experimental economics that will help me to see how this book can help you, which is its purpose.

Study Important Problems

My first experiments were designed to explore questions related to whether, how, and if buyers and sellers in non-durable goods and service markets were well-represented by standard supply and demand (S&D) theory (Smith, 1962). In trying to teach principles I realized that we economists knew nothing about the relationship between S&D and what people do in markets. The trading procedure I used was the oral-outcry two-sided auction, common in securities and commodity markets, because it appeared that those markets were perceived as highly competitive and likely to be good beginning models for testing the theory.

Build on Prior Knowledge

I built upon prior knowledge of the market experiments reported by Chamberlin (1948), an example of which I had participated in as a graduate student. I saw limitations in the design and sought to combine that prior knowledge with independent knowledge of trading procedures used in open outcry markets. Chamberlin was much influenced by Alfred Marshall in applying the concept of reservation price to define willingness to pay (or accept) to buyers (sellers) in the experiment and that carried over into my work. Marshall's modelling of S&D as flows into and out of a consumption market led naturally to repeat trade over time in my design.

What knowledge, how extensive, and when should you acquire and build on it? My style is to do the experimental designs and work, based on the motivating ideas, before examining the literature. Either there is a literature or there is not. If not, it will make little difference. If there is, review the literature when it comes time to write up your work and findings. It is then that you know better what to search for and, more importantly, your work will be independent. Very likely it will be conceived, designed, and motivated differently, constituting a richer contribution that can still be extended in ways suggested by previous results.

Here is a true story on prior knowledge. Roy Radner was working jointly with a probability theorist on a project. They had a draft of a paper finished, and Roy's co-author said he thought it was ready to submit. Roy said, not quite, as they needed to investigate the literature. Roy's co-author said that he would do it, as that would provide him with a good learning opportunity. After a period, the co-author returned saying that this was a remarkable literature. Why? Every paper begins with another paper, not with a problem of the world.

Use Objective Designs

This criterion was satisfied by assigning private values to buyers, and costs to sellers that provided well-defined S&D conditions prior to running each experiment. Eventually, I used other people's classes for subjects besides my own classes and prevailed on others to run experiments. To control for economic understanding, all experiments were run on the first day of economics classes before any text assignments, lectures, or discussion.

I had no experience or background in economics' experiments, as it was not yet a field in economics, but I had studied physics and astronomy, and knew of some of the great experiments in science, giving me a sense of scientific method. Beyond that, common sense seemed to be a good guide.

With those foundations, I designed experiments that could produce findings that would either challenge or support central hypotheses about the operation of markets. For example, competitive equilibria turned out *not* to require large numbers of sellers and buyers.

Provide Full Disclosure

Although I provided complete narrative descriptions of the procedures, it is not clear that this was sufficient for a reader to know how to replicate an experiment. Interestingly, in my second paper the experimental instructions were included in an appendix, so I corrected that error (Smith, 1964).

Use Valid and Reliable Data

For experiments, that means replicate. But none of the original experiments were literally replicated. Motivation for each design and test was followed by only one experiment. Yet the results were startling. Overall, across the experiments, there was a strong tendency to converge to the equilibrium specified by the prior S&D. Hence, the diversity of designs constituted a test of the robustness of equilibrium. The convergence pattern of results, highly variable in terms of convergence paths, stood out prominently relative to variability across experiments.

Use Valid Simple Methods

Simplicity of design allowed the convergence pattern across a diversity of experiments to stand in bold relief. But that outcome was neither anticipated, nor was it an intentional design feature. The experiments were all simple, and the qualitative results transparent in demonstrating convergence. One experiment, with perfectly elastic supply and downward sloping demand failed to converge. It was replicated, using cash payoffs, plus a "commission" to provide a minimum profit for each trade. The replication with cash incentives converged (Smith, 1962, chart 4 and n. 9), which provided early proof of the importance of adequate subject motivation. Subsequently all my experiments paid each subject cash in proportion to profits earned in the experiment.

Objective Designs: Testing Multiple Hypotheses

The first experiments strongly supported convergence. But is there a quantitative rule or law of convergence operating across the experiments? The authors of this book emphasize the importance of testing multiple reasonable hypotheses as the key to achieving objectivity in experiments. Two hypotheses were prominent in the microeconomic literature: (1) Walras postulated that prices rise in proportion to excess demand, fall in proportion to excess supply; (2) Marshall, modeling firm entry, postulated that output from firm entry increased in proportion to the excess of demand price over supply price – a point estimate of profit or loss.

The experiment transaction price histories seemed not to support either (1) or (2); rather a third hypothesis worked much better. Compute $V(p_t)$, the area under the demand, and above the supply curves, at any price p_t). The difference ER = $V(p_t) - V(p^*)$, where p^* is the equilibrium price, appeared to be a better predictor of price at p_{t+1}. Buyers (sellers) were cutting prices to avoid loss from failing to contract. Moreover, V is minimized at p^*, so the process was efficient. New experiment designs, with excess demand constant at all prices p_t, predicted exponential price decay if ER was the rule, constant decay if Walras was right; the data supported ER (Smith, 1965).

Draw Logical Conclusions

The observed pronounced convergence defied prevailing theory, widely shared expectations, beliefs, and teaching. The idea that market participants required either complete information or a Walrasian auctioneer to find prices found no support in the results. Undergraduates, naive in economics, made excellent subjects. Eventually their results generalized across a rich variety of groups (Smith, 1991). Finally, the price paths to equilibrium reflected price concessions by buyers or sellers to avoid failure to make contracts.

Vernon L. Smith

Economic Science Institute
Chapman University
2002 Nobel Laureate in Economics

ACKNOWLEDGMENTS

We thank Joel Kupfersmid, Brian Martin, Frank Schmidt, and Stan Young for reviewing the entire book, and William H. Starbuck for providing useful guidance during the final stages of our book.

Dennis Ahlburg, Hal Arkes, Peter Ayton, Jeff Cai, Nathan Cofnas, John Dawes, John Dunn, Lew Goldberg, Anne-Wil Harzing, Ray Hubbard, Nick Lee, Jay Lehr, Gary Lilien, Byron Sharp, Karl Teigen, and Malcolm Wright reviewed sections of the book that were relevant to their expertise.

Harrison Beard, Len Braitman, Heiner Evanschitzky, Bent Flyvbjerg, Shane Frederick, Gerd Gigerenzer, Andreas Graefe, Jay Koehler, David Legates, Justin Pearson, Don Peters, Paul Sherman, and Arch Woodside made useful suggestions.

We also thank the authors that we cited for substantive findings for checking and improving our summaries of their findings.

Amy Dai and Esther Park – Scott's research assistants – and Charles Green did an excellent job of obtaining and analyzing ratings of papers using the checklists that we have developed for this book, and helped us to improve the clarity of our writing.

Jonathan Ho volunteered to assist us full time during the summer of 2019 and continued to provide help during 2020.

We are grateful to our copy editors, Hester Green, Scheherbano Rafay, Lynn Selhat, and Lisa Zou.

The University of Pennsylvania Library provided much support. Their Document Delivery group was able to track down papers for us, suggest relevant papers, and ensure that our citations were properly formatted.

Scott's wife, Kay Armstrong, has reviewed nearly all of his books and writings over his career. No matter how many reviewers have provided reviews before her, she finds many ways to make further improvements. She has done it again for this book.

WHO IS THIS BOOK FOR?

We wrote this book to help researchers make better use of the scientific method and to write papers and books describing their useful scientific findings in ways that can be understood by the widest relevant audience.

Our book is also intended as a resource for all other stakeholders in science. If you are reading this and are not already a scientist, we expect that you will belong to one or more of the following groups:

- *People who are considering a career as a scientist*, to determine if they are, in fact, well suited for such a career;
- *Employers of scientists*, to help them make hiring decisions;
- *PhD students*, to demonstrate that they can comply with the scientific method;
- *Journal editors*, to increase the publication of papers that comply with the scientific method;
- *Journal reviewers*, to assess the extent to which a paper complies with the scientific method;
- *Government regulatory agencies*, to assess whether current or proposed regulations are, or will be, effective;
- *Policy makers in government or private corporations*, to assess evidence on alternative policies;
- *Courts,* to assess evidence from expert witnesses;

- *Consumers*, to better understand the evidence for product claims, such as when evaluating the efficacy of a medical treatment;
- *Journalists and reporters*, to inform readers about the extent to which a given study complies with science;
- *Citizens*, to analyze evidence for proposed government policies.

AUTHORS' OATH FOR THE SCIENTIFIC METHOD

(1) We, one or both of us, have read each publication that we cite for a substantive finding.

(2) We attempted to contact the authors we cited for substantive findings to help ensure that we accurately described their findings, and to determine if we omitted relevant scientific papers, especially those with scientific evidence that conflicted with our findings.

(3) We used the *Criteria for Compliance with the Scientific Method* from our book, and believe the book is compliant with the scientific method.

(4) Voluntary disclosure: We received no external funding for writing this book and have no conflicts of interest.

1 INTRODUCTION

The scientific method is largely responsible for improving life expectancies and the quality of life over the past 2000 years. Individual scientists, in their efforts to discover how things work and how to make them better have used the method on their own or in collaboration with others to make the world a better place.

We believe that there is no way to improve upon the scientific *method*. Our aim for this book, then, is to help improve scientific *practice*.

The message of our book is a positive one. The scientific method has worked and does continue to work, to the great benefit of all of us. While there are considerable problems with the current practice of science in many fields – which we describe at some length – we provide a plan for overcoming those problems.

Scientists have long been concerned about problems with scientific practice, and leading scientists have long made recommendations for the practice of science. Books by philosophers of science – Karl Popper's *The Logic of Scientific Discovery* (1959; see also Thornton, 2018), and Thomas Kuhn's *The Structure of Scientific Revolutions* (1962; see also Bird, 2018) in particular – also spurred interest in scientific practice.

Then something happened. Michael J. Mahoney published a book – *Scientist as Subject* (1976) – describing the findings of his experimental research on scientific practices. As Mahoney had quickly learned, scientists were not fond of being experimental subjects.

They claimed that he was behaving unethically by not revealing to them that they were participating in an experiment.

Despite the backlash from scientist subjects, the number of papers describing experimental research on scientific practice since Mahoney (1976) has grown at an increasing rate. We draw on those papers to describe failings in current scientific practice, and to provide solutions that are based on scientific research.

Our solutions – recommendations for improving scientific practice so that it complies more with the scientific method – are for scientists, *and* for other stakeholders in the accumulation of scientific knowledge.

Central to our solutions is the *Compliance With Science Checklist* (Chapter 3). All stakeholders can use the checklist. We also provide nine other checklists, one of which is intended for those considering a career as a scientist, and the remaining seven are intended to help practicing scientists with aspects of their role, such as writing a scientific paper.

We believe that the checklists we provide will be useful to all scientists, from PhD students and early career researchers to emeritus professors. We use the term researcher interchangeably with scientist throughout the book.

Before we continue, here is a caution for scientists and science stakeholders, delivered in the Nobel Prize lecture by Friedrich von Hayek:

> Yet the confidence in the unlimited power of science is only too often based on a false belief that the scientific method consists in the application of a ready-made technique, or in imitating the form rather than the substance of scientific procedure, as if one needed only to follow some cooking recipes to solve all social problems. It sometimes almost seems as if the techniques of science were more easily learnt than the thinking that shows us what the problems are and how to approach them. F. A. Hayek (1974)

1.1 Plan of This Book

We survey the current state of scientific practice in Chapters 4 through 7. Chapter 4 (Assessing the Quality of Scientific Practice) describes how we reviewed the evidence on practice and sets the scene. Chapter 5 examines the problem of advocacy, Chapter 6 the problem of mandatory journal peer reviews, and Chapter 7 the problems created by government funding and regulating research.

In Chapter 8, we describe research findings on what it takes to be a good and useful scientist.

We then describe solutions in Chapters 9–11. Chapter 9 provides guidance for scientists on how to discover useful scientific knowledge. Chapter 10 provides guidance on how to write a scientific paper to best communicate useful research findings, and on how to disseminate those findings to the widest relevant audience. Chapter 11 provides guidance on how stakeholders can help science in four sections devoted to university managers, scientific journal editors, governments and courts, and media and interested individuals respectively.

Chapter 12 summarizes how our checklists and guidance provide a plan for reforming scientific practice so that the practice of science matches the scientific method. We expect great benefits to flow from reducing unscientific practices and increasing the adoption of the scientific method throughout the research process from the production to the consumption of useful scientific knowledge.

Finally, an Afterword by Terence Kealey – author of *The Economic Laws of Scientific Research* (1996) – provides a fascinating account of the history of science that revolves around the question "How do we know this statement is true?" He reinforces and expands on the conclusion of this book that the increasing involvement of governments in research from the mid-twentieth century has diverted science from its true role as an engine for discovering useful truths, and endorses our recommendations for returning the scientific endeavor to its true path.

1.2   Scientific Method versus Scientist Opinion

The opinions of scientists – even those of the most eminent – should not be confused with knowledge obtained from the application of the scientific method. Scientists' opinions are examples of the logical fallacies of appeal to authority and – when the opinion is shared by a group of scientists – *argumentum ad populum*.

Not surprisingly, then, scientists' opinions about the way things are or will be have not held up well against reality. Cerf and

Navasky's (1998) collection of scientists' and other widely respected experts' opinions, *The Experts Speak: A Definitive Compendium of Authoritative Misinformation*, is a thick book (445 pages) with many examples. Here is one: "Heavier-than-air flying machines are impossible" (Lord Kelvin, British mathematician, physicist, and President of the British Royal Society, 1895.)

Some of the Cerf and Navasky examples are hilarious, if one ignores the harm that they caused when they were taken seriously. There is an abundance of what seem now to have been idiotic prognostications, supporting George Orwell's observation that "One has to belong to the intelligentsia to believe things like that: no ordinary man could be such a fool" (Orwell, 1945, para 29). The book was sufficiently successful that the authors were encouraged to put out a revised edition in 1998 with more examples of incorrect, but highly confident, expert opinions.

An experiment to assess the value of expert judgments was conducted over a 20-year period (Tetlock, 2005). The 284 experts who participated were asked to assess the probabilities of various events occurring for situations in the future. The experts were people whose professions included "commenting or offering advice on political and economic trends." By 2003, Tetlock had accumulated 82,361 forecasts. He then evaluated the experts' judgments against the outcomes, and against predictions from simple statistical procedures, uninformed non-experts, and well-informed non-experts. The experts barely, if at all, outperformed the informed non-experts and none of the groups did well against simple rules and models. (Tschoegl and Armstrong, 2007, reviewed the book.)

Scott had previously reviewed experimental evidence on experts' judgmental predictions. The review led him to develop his Seer-Sucker Theory: "No matter how much evidence exists that seers do not exist, suckers will pay for the existence of seers" (Armstrong, 1980a).

When knowledge about a situation is at best tentative, scientists nevertheless can and do use their perceived authority to promote theories in the hope of persuading voters, government officials, and political leaders that there is a problem and that government actions that accord with their opinions are needed. The approach, which is an embodiment of the "precautionary principle," has been called "post-normal science" (Ravetz, 2004). The precautionary principle is an anti-scientific political principle that, in the absence of objective cost-benefit analyses, is used to call for drastic government actions in response to some scientists' opinions that bad things will happen otherwise (Green and Armstrong, 2008).

1.3   Objective of the Scientific Method

Benjamin Franklin believed that universities should be centers for scientific research. When he founded what is now known as the University of Pennsylvania, he suggested that faculty be involved in the "discovery and dissemination of *useful* knowledge" (Franklin, 1743). We believe that his suggestions should be the objective of all scientific research, and the yardstick against which it is judged.

Other pioneers of science professed a similar preference for usefulness or importance, as we discuss in Chapter 2 (Defining the Scientific Method). But while the scientific method is efficient for making useful discoveries – because it is designed to identify the hypotheses that best accord with reality – it is up to scientists to identify the problems that are most likely to lead to useful discoveries, and that they can best help with.

1.4   Objectives of Scientific Practice Subverting Science

Outside of the business world, current procedures for the evaluation of researchers' contributions provide little to encourage them to achieve the objective of discovering useful knowledge. Instead of assessing the usefulness of scientists' research findings, their employers use proxy measures such as the number of papers published in "high-quality" journals, citation counts, and dollars of grant money received.

The outcome of that approach is consistent with Campbell's Law: "The more any quantitative measure is used for social decision-making, the more subject it will be to corruption pressures and the more apt it will be to distort and corrupt the social processes it is intended to monitor" (Campbell, 1979, p. 85). For example, researchers are motivated to divide a research project into a series of papers, and to include as co-authors people with little or no involvement in the research project on the understanding that the favor will be reciprocated. *The Economist* (2016) examined more than 34,000 papers listed on Scopus between 1996 and 2015 and found that the average number of authors per paper grew from 3.2 to 4.4.

While the proxy measures may have been reliable indicators of useful scientific findings when they were first adopted, they are no longer so.

Citation analysis began in 1961 when Eugene Garfield began publication of the Science Citation Index (SCI) in Philadelphia. The Social Science Citation Index (SSCI) followed. These provide a valuable service for scientists who are searching for prior knowledge in their area of study.

Since then, there has been a phenomenal increase in the number of citations in all fields of science. For example, when Scott joined The Wharton School in 1968, the author with the most citations in the field of marketing, Paul Green, received 100 citations in some years. He was one of the most renowned researchers in marketing of his time. Things are vastly different now. As at September 2019, one researcher in marketing had received more than 15,000 citations in one year.

The number of citations that a paper receives *might* provide an indicator that it has no scientific value when other scientists fail to cite it. But even then, a paper that challenges the current orthodoxy in a field might be ignored regardless of its contribution to scientific knowledge.

The vast majority of papers are not cited in any substantive manner. For example, Armstrong and Overton (1977) developed a simple and effective way to estimate non-response bias in mail surveys. The paper had been cited more than 15,000 times as of early 2020 according to Google Scholar. Yet, in a sample of 50 papers in the leading journals that cited the Armstrong and Overton procedure, only one correctly represented the procedure. Most of the citations used it to support their own incorrect procedures for dealing with non-response bias, suggesting that the authors of those papers had not even read the Armstrong and Overton paper. In addition, many of the citations made mistakes in the references, such as incorrect spellings, which were often identical to the mistakes made by other authors who had cited the paper (Wright and Armstrong, 2008).

Do those who cite papers regard them as useful? Apparently not. An analysis of 12 high-profile scientific papers estimated that about 70–90 percent of cited papers had not been read by those who cited them (Simkin and Roychowdhury, 2005).

Fire and Guestrin (2019) analyzed more than 120 million papers in 2,600 fields, with an emphasis on the field of biology. They concluded that, "citations are not beneficial for comparing researchers ... even in the same department" (p. 76).

Citations of papers that have been refuted or challenged, or retracted, contribute to the authors' citation counts even though they are detrimental to science. For example, a study of biomedical papers from 1966 through August 1997 found that 235 papers had been retracted. These retracted papers were cited 2,034 times, *after being retracted*. On all but 19 occasions, the citation was treated as a valid study (Budd et al., 1998).

Another problem is that many are "mysterious citations." That is, the authors do not explain what findings the papers are being cited for, nor how they were discovered. We provide guidance on avoiding mysterious and unnecessary citations in Chapter 10 (How Scientists Can Disseminate Useful Findings) and Chapter 11 (How Stakeholders Can Help Science).

We discuss the corrupting incentives resulting from the involvement of governments in research via grant funding and regulation in Chapter 7 (Scientific Practice: Problem of Government Involvement), and solutions to that problem in Chapter 11 (How Stakeholders Can Help Science).

1.5   Operational Guidelines for Scientific Practice

In this book, we develop guidelines for implementing the scientific method. To do that, we first derived a list of criteria for complying with the scientific method from descriptions provided by the founders of the method. We then translated those criteria into a checklist of operational guidelines that can be used to determine the extent to which a research paper complies with the scientific method.

We argue that the criteria should be generally acceptable given that they are based on those proposed by the scientists who originated the scientific method, and the success in generating useful knowledge that following the criteria has had across 21 centuries.

1.5.1   Previous Attempts at Guidelines for Science

In the past, various disciplines developed guidelines for scientific practice based on a consensus of the opinions of scientists working in the field. An early instance was the Operations Research Society of America's "Guidelines by the Ad Hoc Committee on Professional Standards" (ORSA, 1971). Guidelines have also been developed for medical research and clinical practice, including GRADE (Guyatt et al., 2008), CONSORT (Moher et al., 2010; Schulz et al., 2010), and SQUIRE (Davidoff et al., 2008).

For 70 years, the US Supreme Court followed the Frye standard in assessing scientific evidence. That standard required courts to follow the "generally accepted opinions of scientists" – another consensus-based standard.

The court's consensus standard changed following the 1993 *Daubert* v. *Merrell Dow* case, in which the Supreme Court of the United States unanimously replaced the Frye approach in favor of assessing evidence on the basis of whether it was the product of "scientific procedures." To date, about half of US state courts have adopted the Daubert approach.

To help the courts implement the Daubert standard, descriptions of scientific procedures have been distributed to all federal judges in the *Reference Manual on Scientific Evidence* (Breyer et al., 2011). By 2011, the third edition contained over one thousand pages. The Daubert standard has had great effect on the legal system in the United States and, according to some lawyers, has led to better judgments (e.g., Faigman, 2013).

The Daubert standard does, however, depend upon the authority of selected experts to choose the proper procedures. In addition, consensus on what are the appropriate scientific procedures may change over time as better procedures are discovered. Scientists must, therefore, keep up to date with the development of procedures and evidence on their validity.

1.5.2 Guidelines Necessary, but Not Sufficient

The development of guidelines, while necessary, is not sufficient, as a number of studies have found. One study examined six different sets of guidelines for medical research: The authors concluded that the "implementation of these guidelines has led to only a moderate improvement in the quality of the reporting of medical research" (Johansen and Thomsen, 2016).

1.5.3 Mandated Checklists Necessary

The only way we know to ensure compliance with guidelines is the required and monitored use of an operational checklist of the guidelines.

The effectiveness of monitored checklists has been well documented. For example, a review of 15 experimental studies in health care found that validated checklists led to substantial improvements in patient outcomes. One of the experiments examined the application of a 19-item checklist for a surgical procedure that was performed on thousands of patients in eight hospitals around the world. Use of the checklist reduced mortality rates at those hospitals by half (Haynes et al., 2009).

Checklists are especially effective when people know little about the relevant scientific principles. For example, advertising novices were asked to use a checklist with 195 validated persuasion principles to rate 96 pairs of advertisements. By using the checklist, they made 44 percent fewer errors than did unaided novices in predicting which advertisements were more effective (Armstrong et al., 2016).

Checklists also help when the users *are* aware of proper procedures. For example, an experiment on infection prevention in the intensive care units of 103 Michigan hospitals required physicians to follow five *well-known* guidelines for inserting catheters. Use of the checklist reduced infection rates from 2.7 per 1,000 patients, to zero after three months (Hales and Pronovost, 2006).

Users should confirm that they have implemented each item of a comprehensive checklist, and the use of the checklist should be monitored. In some fields, such as engineering, aeronautics, and medicine, failure to follow *operational, agreed-upon* checklists can be used by courts to assign blame for bad outcomes. In some cases, the failure to complete a checklist can be grounds for dismissal of an employee.

2 DEFINING THE SCIENTIFIC METHOD

The invitation for those nominating candidates for the Nobel Prize in economics, the "Sveriges Riksbank Prize in Economic Sciences in Memory of Alfred Nobel," described the award of the prize as being "based solely on scientific merit." No criteria for judging scientific merit were provided, but nominators were directed to "consider origin and gender" of the nominees. Without clear criteria for the award, to what extent can one be confident that the prize was based on the scientific merit of the findings?

In this chapter we provide an aspirational definition of the scientific method. The definition is in the form of eight criteria that are based on the writings of key figures in the development of the scientific method. We then expand on each of the criteria, describing their source – where appropriate – and the reasons for their importance for the scientific method.

2.1   An Aspirational Definition

We sought to define the scientific method in such a way that most researchers *should* aspire to the ideal the definition represents. To do so, we turned to the writings of the developers of the scientific method. Scientists have been describing elements of the scientific method since before 400 BC. White (2002) concluded that the modern scientific method owes its approach to the logical framework of hypothesis testing laid out by Socrates, with later refinements by Plato and Aristotle. Socrates in effect set out the basis of a valid approach to seeking knowledge that scientists still use – the use of experiments, which came to be formally recognized as important much later, excepted.

We concluded that the key elements of the scientific method – as derived from the words of famous and pioneering scientists – could be summarized by eight criteria:

1. Study important problems
2. Build on prior knowledge
3. Provide full disclosure
4. Use objective designs
5. Use valid and reliable data
6. Use valid simple methods
7. Use experimental evidence
8. Draw logical conclusions

These criteria are also consistent with the *Oxford English Dictionary (OED)*, which defines the scientific method as:

> commonly represented as ideally comprising some or all of (a) systematic observation, measurement, and experimentation, (b) induction and the formulation of hypotheses, (c) the making of deductions from the hypotheses, (d) the experimental testing of the deductions, and (if necessary) (e) the modification of the hypotheses … The modern scientific method is often seen as deriving ultimately from Francis Bacon's Novum Organum (1620) and the work of Descartes. In the 20th century, Karl Popper's idea of empirical falsification has been important. OED Online (2018).

In practice, a study can *contribute* to making a useful scientific discovery even when it does not on its own comply with all of the criteria. For example, Einstein drew on the findings of others' experiments to develop novel hypotheses about important problems that could in turn be tested against alternative hypotheses by further experiments.

Papers might also contribute to science by identifying important problems. Others might contribute by identifying shortcomings in the papers of other researchers and resolving those issues. Another contribution is to develop objective measures of important variables, and compile data using those measures, as has been done by scientists at the University of Alabama at Huntsville in estimating global average temperatures from satellite readings (Spencer et al., 2017).

While studies that fall short on some criteria – e.g., by over-looking prior knowledge – might nevertheless turn out to provide a useful contribution to research on a problem, studies that failed to use an objective design (criterion 4) are unlikely to do so. In order to claim that a principle or method is scientific, studies of the problem would, when taken together, need to satisfy all eight criteria.

We consider that the support of meta-analyses of *objective* studies that collectively comply with *all eight criteria for science* are necessary for rational policy making. The requirement is particularly important for government laws and regulations, because they involve duress rather than voluntary transactions.

2.2   Criteria for Complying with the Scientific Method

We now expand on the eight criteria for complying with the scientific method that we described above.

2.2.1   Study Important Problems

> According to the general spirit of this book, which values everything in its relation to Life, knowledge which is altogether inapplicable to the future is nugatory.
>   Charles Sanders Peirce
>   (1958, para 56)

Scientists in the past sought to address important problems. Robert Boyle, a founder of the English Royal Society, wrote in 1646 that the founders valued "no knowledge but that it has a tendency to use" (as quoted by O'Connor and Robertson, 2004).

Some scientists argue that research that does not obviously lead to useful findings is nevertheless important because of potential future useful-ness. While that may turn out to be true in some cases, identifying problems that are currently in need of solutions to research is more likely to produce useful findings than is research based on curiosity about a non-problem.

Addressing currently pressing problems can lead, and has led, to advances in scientific knowledge that go well beyond finding solu-tions to those problems, as the following quotation illustrates.

> [T]he practical sciences incessantly egg on researches into theory. For considerable parts of chemical discovery we have

to thank the desire to find a substitute for quinine or to make quinine itself synthetically, to obtain novel and brilliant dye-stuffs, and the like. The mechanical theory of heat grew out of the difficulties of steam navigation. For it was first broached by Rankine while he was studying how best to design marine engines. Then again, one group of scientists sometimes urges some overlooked phenomenon upon the attention of another group. It was a botanist who called van't Hoff's attention to the dependence of the pressure of sap in plants upon the strength of the solution, and thus almost instantaneously gave a tremendous impulse to physical chemistry. In 1820, Kästner, a manufacturer of cream of tartar in Mulhouse, called the attention of chemists to the occasional, though rare, occurrence in the wine casks of a modification of tartaric acid, since named racemic acid; and from the impulse so given has resulted a most important doctrine of chemistry, that of the unsymmetric carbon atom, as well as the chief discoveries of Pasteur, with their far-reaching blessings to the human species. Charles Sanders Peirce (1958, para 52)

If research on relatively narrow current problems can lead the curious scientist to such widely important discoveries as are described in the quotation from Peirce (1958) above, the case for studying non-problems at someone else's expense seems weak when researcher time is a limited resource. Of course, if there is a willing well-informed funder for such activity, including self-funding, then that is the business of the parties concerned, and good luck to them.

2.2.2   Build on Prior Knowledge

Progress in science requires that scientists become familiar with prior knowledge and methods for the given problem. Newton (1675) referred to the process as "standing on the shoulders of giants."

Despite the logical necessity of doing so, researchers often fail to comprehensively review the existing evidence, perhaps because doing so greatly increases the time needed to complete a publication. Because the reviewers used by journal editors are often unaware of relevant prior scientific findings, an author's failure to identify relevant prior research can go undetected. As a consequence, researchers are prone to making *re*discoveries.

In one example, Kahneman (2011) concluded that people process information differently depending on the nature of the decision. He referred to the phenomenon as "slow versus fast," or "System 1" and "System 2" decision-making. His was at least the third discovery of the concept. In 1913 it was called "short circuit versus long-circuit" thinking as described by Hollingworth (1913). Half-a-century later, the concept was referred to as "low involvement versus high-involvement" by Krugman (1965). Whatever name the concept is given, it has been an important condition to consider for persuasion for over a century now. For more on this, see Armstrong (2010, pp. 21–22).

2.2.3   Provide Full Disclosure

The scientific method depends heavily on replication, and replication requires full disclosure of methods. Replications are needed to help determine whether potentially useful scientific findings should be accepted and acted upon.

A paper that does not provide all necessary information for replication may, nevertheless, contribute to science if it at least addresses an important problem. Other researchers can conduct *extensions* that test the same issue. The extensions can help to allay concerns about findings that arise when disclosure is incomplete.

2.2.4   Use Objective Designs

The founders whose writings we used to develop the definition of the scientific method recognized early on that objectivity is hard to achieve. They also recommended a solution. Sir Isaac Newton, for example, described four "Rules of Reasoning in Philosophy" in the third edition of his Philosophiae Naturalis Principia Mathematica (1726, pp. 387–389). His fourth rule, in Motte's translation from Latin, states, "In experimental philosophy we are to look upon propositions collected by general induction from phenomena as accurately or very nearly true, *notwithstanding any contrary hypotheses that may be imagined*, till such time as other phenomena occur, by which they may either be made more accurate, or liable to exceptions" (Newton, 1729, vol. 2, p. 205, emphasis added).

We refer to this solution as Multiple Reasonable Hypotheses Testing, or MRHT. One should include all reasonable hypotheses or describe why that was not feasible. MRHT stands in contrast to the approach that has become accepted practice in psychology and the social sciences: Null Hypothesis Statistical Testing, or NHST.

The increase in productivity that arose from the English Agricultural Revolution illustrates the importance of MRHT. Agricultural productivity saw little improvement until landowners in the 1700s began to conduct experiments comparing the effects of alternative ways of growing crops. The Industrial Revolution progressed in the same manner (Kealey, 1996, pp. 47–89).

Chamberlin (1890) claimed that disciplines that conduct experiments to test multiple reasonable hypotheses progress greatly, while those that do not, progress little. Nearly three-quarters of a century later, Platt (1964) reiterated Chamberlin's conclusion because researchers in many fields of science were still ignoring the original advice.

MHRT has also led to advances in medical knowledge. For example, one study examined all papers that used MRHT that were published in the *New England Journal of Medicine* from 2001 to 2010 (Prasad et al., 2013). The study found that 146 medical treatment recommendations were reversed as a consequence of experiments using MRHT. The reversals amounted to 40 percent of all procedures tested. MRHT has also led to the growth of useful knowledge in engineering, forecasting, persuasion, and technology.

2.2.5   Use Valid and Reliable Data

Validity is the extent to which the data measure the concept that they purport to measure. Validity is not a trivial matter. Many disputes arise due to differences in how concepts are measured. For example, what is the best way to measure inequality among people? Is it best assessed only in terms of money income, or should it also include the effects of taxes, wealth, transfer payments, home production, etc.? These measures produce different findings and policies. More fundamentally, should inequality be assessed in terms of life satisfaction instead of income? Money income is, after all, only one of several means to achieve the desired end of happiness. People routinely trade off money income to do work that provides greater intrinsic satisfaction or to live somewhere that they prefer.

Reliability is established when other researchers, using the same procedures, can reproduce findings. Reliability can be improved by using all relevant data that are available such as when using a time-series. As Sir Winston Churchill said, "The longer you can look back, the farther you can look forward."

Data that has been subject to unexplained revisions should not be used. Enough said.

2.2.6   Use Valid Simple Methods

> There is, perhaps, no beguilement more insidious and dangerous than an elaborate and elegant mathematical process built upon unfortified premises.
> Chamberlin (1899, p. 890)

Validity requires that the method used has been tested and found to be useful for the problem at hand. Simple methods are those that can be understood by those who might have an interest in reading or replicating the paper. Complex methods make it difficult for others to understand the research, spot errors, and replicate the study.

The call for simplicity in science started with Aristotle but is usually attributed to Occam as "Occam's Razor." Yet, academics and consultants love complex methods. So do their clients. After all, if the process were simple they would ask, "Why are we paying all that money?" For a further discussion of why complexity proliferates, see Hogarth (2012).

The 1976 Nobel Laureate in Economics, Milton Friedman, stressed the importance of testing the predictive validity of hypotheses against new, or out-of-sample, observations (1953). Is there a conflict between predictive validity and simplicity? Apparently not. Comparative studies have shown the superior predictive validity of simple methods in out-of-sample tests across diverse problems. The experiments on the predictive validity of simple alternatives to multiple regression analysis by Czerlinski et al. (1999), and by Gigerenzer et al. (1999) are elegant examples.

In our review of the evidence on the predictive validity of Occam's Razor, we defined a "simple method" as one for which an intelligent person could understand: (a) procedures; (b) representation of prior knowledge; (c) relationships among the elements; and (d) relationships among models, predictions, and the decisions that might be made (Green and Armstrong, 2015). We found 32 published studies that compared forecasts from simple methods with forecasts from more complex methods that had been proposed by their authors as a way to improve accuracy. We hired university students to rate complexity against the simplicity criteria listed above. Simplicity improved out-of-sample predictive validity in all 32 studies involving 97 experimental comparisons. On average, complex methods had errors for out-of-sample predictions that were 27 percent larger for the 25 papers that provided quantitative comparisons. The strength and consistency of the findings astonished us and are a caution to researchers who assume that complex data modelling methods have predictive validity.

2.2.7 Use Experimental Evidence

> The testing of the hypothesis proceeds by deducing from it experimental consequences almost incredible, and finding that they really happen, or that some modification of the theory is required, or else that it must be entirely abandoned.

> These experiments need not be experiments in the narrow and technical sense, involving considerable preparation. That preparation may be as simple as it may. The essential thing is that it shall not be known beforehand … how these experiments will turn out. Charles Sanders Peirce (1958, paras 83, 90)

Experiments emerged as a key element of the scientific method in the practice of the natural sciences in the sixteenth century. The importance of experiments was generally not recognized in medical research and the social sciences until the nineteenth century (DiNardo, 2018).

Robert Boyle and other scientists established the forerunner of the modern-day Royal Society around 1645 to acquire knowledge through experiments. The value the society placed on experiments was highlighted by the appointment of Robert Hooke as a Curator of Experiments who was tasked with "furnish[ing] them every day on which they met with three or four considerable experiments" (O'Connor and Robertson, 2004). The society translates its Latin motto, *nullius in verba*, as "take nobody's word for it." It expresses the Royal Society Fellows' determination "to withstand domination of authority and to verify all statements by an appeal to facts determined by experiment" (Royal Society, 2019).

Experiments can be controlled, quasi-controlled – include some, but not all, important causal variables – or natural. Laboratory experiments allow for more control over conditions, while field experiments are more realistic. Interestingly, a comparison of findings from laboratory versus field experiments in 14 areas of organizational behavior concluded that they produced similar findings (Locke, 1986). Vernon Smith demonstrated that "laboratory" (controlled) experiments can be used to test competing hypotheses in economics. He found that very simple experiments could be devised that would replicate the relevant behaviors of participants in real markets (Smith, 2002).

Experiments have been conducted in fields of science as diverse as astronomy (e.g. Ostro, 1993, described the use of radar to conduct experiments on the scale of the solar system and gravitation, among other things), evolutionary biology (e.g., Schluter, 1994, conducted experiments to test theories about the effect of resource competition among species on evolution), geology (Kuenen, 1958, described the use of experiments in geology starting with those of Sir James Hall, who began conducting his experiments in 1790), paleontology (e.g., Oehler, 1976, described experiments that simulated fossilization in synthetic chert), and zoology (e.g., Erlingsson, 2009, described the rise of experimental zoology in Britain during the 1920s).

Darwin is most famous for his theory of evolution, but he also devoted much time to testing hypotheses with experiments. For example, he hypothesized, contrary to then current belief, that plants move, and designed experiments that tracked plant movement (Hangarter, 2000). But not all research problems are amenable to testing by way of experiments that are controlled by the researcher, as Mayr (1997) described in his book on the science of biology: "Much progress in the observational sciences is due to the genius of those who have discovered, critically evaluated, and compared … natural experiments in fields where a laboratory experiment is impractical, if not impossible" (p. 29).

Natural experiments have been used to test competing theories in the physical sciences; for example, Maupertuis's expedition to Lapland over the winter of 1736–1737 to undertake observations that would test the Cartesian theory that the earth is taller than it is broad against Newton's theory that the opposite is the case. More famously, Eddington's 1919 expeditions were mounted to determine whether Einstein's or Newton's gravitation theories provided the better prediction of phenomena by taking advantage of the natural experiment provided by a solar eclipse (Sponsel, 2002).

Hypotheses on the distribution of plants from Darwin's speculations and findings from experiments on the survival and dispersal of plant seeds (Carlquist, 2009) were tested by the natural experiment of the 1883 eruption of the island of Krakatoa (Krakatau). The eruption sterilized what was left of the island such that most plant life – with the possible exception of some grasses – would have to have arrived on or over open sea. Nine months after the eruption, there was no sign of plant life, but by 1930 the whole island was covered with dense forest (Went, 1949).

Gould (1970) advocated greater use of experiments in paleontology – "we must include the experimental approach ... and not remain tied to the observational mode of traditional natural history" (p. 88) – and described prior studies that used natural experiments. He quoted Seilacher on the topic: "One cannot make experiments with organisms that became extinct hundreds of million years ago. Still, isn't it an experimental approach if the belemnites' habits were tested through the reactions of its commensals? The fact that the actual test was made long before man's existence does not alter the principles of its evaluation" (Gould, 1970, p. 89).

Variations between the societies of different countries, regions, states, and communities, and changes over time provide natural experiments against which researchers can test hypotheses from alternative theories. Diamond and Robinson's (2010) edited book *Natural Experiments of History* includes seven analyses of political and social arrangements and their economic outcomes or causes using natural experiments from history. Alternative arrangements for managing common pool resources provide natural experiments that allowed testing of hypotheses on whether sustainable management arrangements can arise by trial and error, or whether they must be imposed by a political authority (Ostrom, 1990). Variations in regulations between US counties and states, and over time, allowed Lott (2010) to test hypotheses on the relationship between gun control and crime.

Note that some scientists consider the term "natural experiments" to be only a metaphor for studies that literally test hypotheses by making observations, or "observational studies," and not *true* experiments. We prefer to use the term "natural experiments" in order to distinguish studies that are properly designed to test alternative hypotheses by identifying situations in which observations might turn out to falsify them, and reserve the term "observational studies" for studies that do not test hypotheses or that develop hypotheses to fit observations.

For ideas and guidance on designing experiments see Shadish, Cook, and Campbell's (2001) book *Experimental and Quasi-Experimental Designs for Generalized Causal Inference*. They describe diverse and creative ways to conduct experiments. Another resource is Dunning's (2012) book *Natural Experiments in the Social Sciences: A Design-Based Approach*, the first part of which is devoted to "discovering natural experiments."

Experiments guided by sound theoretical reasoning provide the only valid and reliable way to establish *causal relationships*. Causality cannot be identified by "machine learning" methods, known by names such as artificial intelligence, data mining, factor analysis, and stepwise regression. We described the lack of evidence that the models that are the product of machine learning methods have any predicted validity in our 2018 and 2019 co-authored papers.

Machine learning models violate the scientific method because they fail to incorporate prior knowledge from experimental studies and coherent theory. The models are also vulnerable to including variables that have no known causal relationship to the variable of interest. As economist Friedrich Hayek warned in his Nobel Prize lecture, "in economics and other disciplines that deal with essentially complex phenomena, the aspects of the events to be accounted for about which we can get quantitative data are necessarily limited and may not include the important ones" (Hayek, 1974).

Meta-analyses of experimental data are the gold standard of evidence. Meta-analyses combine the results of all experimental studies on the issue being studied, no matter the type of experiment. For example, a meta-analysis of 40 experiments on how communication affects persuasion found the conclusions from field and laboratory studies were similar (Wilson and Sherrell, 1993).

Findings from experimental studies do, however, often differ from those based on non-experimental data. For example, expert judgments and non-experimental research typically conclude that consumer satisfaction surveys improve consumer satisfaction. However, well-designed experiments showed that they *harm satisfaction* because customers look for bad things to report. They also create dissatisfaction among those providing the services. The problems went away when people were asked *what they liked about the product or service* (Ofir and Simonson, 2001).

Non-experimental data from hundreds of thousands of users showed that female hormone-replacement therapy helped to preserve youth and ward off a variety of diseases in older women. The findings were replicated. However, subsequent experimental studies found that the treatment could actually be harmful. The favorable findings from the non-experimental data occurred because the women who used the new medicine were generally more concerned about their health and sought out ways to stay healthy (Avorn, 2004).

Kabat's (2008) book on environmental hazards – examining such topics as DDT, electromagnetic fields from power lines, radon, and second-hand smoke – concluded that analysis of non-experimental data in studies on health had often misled researchers, doctors, patients, and the public.

Non-experimental data analyses lend themselves to advocacy studies. They allow researchers to produce "evidence" for almost any hypothesis by attributing causal relationships to correlations in survey data.

Vernon Smith, a pioneer of experimental economics and a 2002 Nobel Laureate in Economics, suggested that what can be learned from well-designed laboratory experiments is only limited by the ingenuity and creativity of the researcher.

> What are the limits of laboratory investigation? I think any attempt to define such limits is very likely to be bridged by the subsequent ingenuity and creativity … of some experimentalist. Twenty-five years ago I could not have imagined being able to do the kinds of experiments that today have become routine in our laboratories. Experimentalists also include many of us who see no clear border separating the lab and the field. Vernon Smith (2003, p. 474, n. 27).

There may be problems or situations for which experiments are not possible. In such cases, analyses of non-experimental data may be useful for helping to identify *whether hypothesized causal relationships are plausible*. For situations in which causal relationships have been established, analyses of non-experimental data can help to assess effect sizes.

Some philosophers of science have theorized that experiments cannot do what scientists expect them to: contribute to knowledge by rejecting or supporting hypotheses. As we hope is clear from this book, we disagree, strongly. Philosopher of science Deborah Mayo and practitioner of science Vernon Smith have also disagreed, as follows.

> In principle the D-Q problem[1] is a barrier to any defensible notion of a rational science that selects theories by a logical process of confrontation with scientific evidence. This is cause for joy not despair. Think how dull would be a life of science if, once we were trained, all we had to do was to turn on the threshing machine of science, feed it the facts and send its output to the printer. In practice the D-Q problem is not a barrier to resolving ambiguity in interpreting test results. The action is always in imaginative new tests and the conversation it stimulates. My personal experience as an experimental economist since 1956, resonates well with Mayo's critique of Lakatos:
>
> Lakatos, recall, gives up on justifying control; at best we decide – by appeal to convention – that the experiment is controlled ... I reject Lakatos and others' apprehension about experimental control. Happily, the image of experimental testing that gives these philosophers cold feet bears little resemblance to actual experimental learning. Literal control is not needed to correctly attribute experimental results (whether to affirm or deny a hypothesis). Enough experimental knowledge will do. Nor need it be assured that the various factors in the experimental context have no influence on the result in question – far from it. A more typical strategy is to learn enough about the type and extent of their influences and then estimate their likely effects in the given experiment. Vernon Smith (2002, p. 106, quoting Mayo, 1996, p. 240)

2.2.8   Draw Logical Conclusions

Francis Bacon (1620 [1863]) reinforced Aristotle's assertion that the scientific method involves logical induction from systematic

---

[1] The Duhem-Quine problem is the assertion that designing an experiment to test a hypothesis is not possible without making assumptions or involving additional hypotheses that may themselves be the cause of the experiment's support for or rejection of the hypothesis (the authors).

observation. Conclusions should follow logically from the evidence provided in a paper.

How might logic be used to compare competing hypotheses? Here is an example: compare the hypothesis that people in a given community in a rich country will be happier if the government redistributes money income from higher income people to those with lower incomes (Hypothesis #1), with the hypothesis that people in a community who are happier are more productive and earn more money (Hypothesis #2), and with the hypothesis that the happiness of people within a community is more affected by their relative status than by their absolute money income (Hypothesis #3). The latter hypotheses lead to policy conclusions that are opposite to the those from the first. Frey's (2018) summary of evidence from happiness research provides support for Hypothesis #2 and #3, and cautions against Hypothesis #1.

If the research addresses a problem that involves strong emotions, consider writing the conclusions using symbols in order to check the logic. For example, the argument "if P, then Q. Not P, therefore not Q" is easily recognized as a logical fallacy – known as "denying the antecedent" – but recognition is not easy for contentious issues, such as the relationship between guns and crime.

Violations of logic are common in the social sciences. We suggest asking researchers who have different views on the problem you are studying to check your logic. Logic does not change over time, nor does it differ by field. Thus, Beardsley's (1950) *Practical Logic* continues to be useful. For an additional discussion of logical fallacies, see the website www.logicalfallacies.org.

3 CHECKLIST FOR THE SCIENTIFIC METHOD

We intend that our checklist provides a common understanding among all stakeholders in science of what the scientific method entails. To that end, we describe it in terms that are simple and commonly understood.

In this chapter, we outline how we developed the *Compliance with Science Checklist*. We then present the checklist of eight criteria for complying with the scientific method and 26 items to help check whether the criteria are met. The checklist is intended for all stakeholders of science. We describe how the checklist can be used, and list stakeholders and what they can use the checklist for in Table 3.1. We caution that checklists are only useful if they are logical and based on evidence, and if they are used.

3.1 Development of the *Compliance With Science Checklist*

Checklists draw upon the decomposition principle, which reduces a complex problem into simpler parts. One solves or makes estimates for, or rates, each part, and then calculates an aggregate solution, or overall rating.

Our review of experimental evidence showed that decomposition typically provides substantial improvements in predictive validity. For example, in three experiments on subjects' decisions for job and college selection, judgmental decomposition resulted in more accurate judgments than holistic ratings (Arkes et al., 2010). Similarly, an experiment in which members of the Society for Medical Decision Making

Table 3.1. Potential users and uses of the *Compliance With Science Checklist*

Researchers	• determining which findings to cite • ensuring that their own papers comply • informing clients, editors, users, and readers on the extent to which their paper complies
Journals	• setting expectations of authors • identifying which criteria were met • selecting which papers to publish
Universities	• training, hiring, promoting, and dismissing scientists • setting expectations of researchers • disseminating useful scientific findings
Think Tanks	• assessing papers to identify the scientific criteria that were met
Funders	• requiring research to meet scientific criteria
Awards Committees	• choosing recipients who made useful scientific discoveries
Certifiers	• independently assessing the extent to which papers provide useful scientific findings
Managers	• assessing the value of published findings
Journalists	• reporting the extent to which studies address important problems and comply with science
Regulators	• developing, revising, and rescinding regulations based on compliance with science
Law Courts	• assessing the value of evidence

evaluated presentations at their annual convention found that decomposed ratings were more reliable than holistic ratings (Arkes et al., 2006). For additional experimental evidence on the value of decomposition, see MacGregor (2001).

To develop a checklist of criteria for compliance with the scientific method, we reviewed experimental research on scientific practice (described in Chapter 4). Based on the research findings, we designed *operational guidelines* for each of the eight criteria. For example, to gauge a paper's objectivity, the checklist asks raters to determine whether a paper compares multiple reasonable hypotheses.

As we will show in this book, the *Compliance With Science Checklist* provides a valid and reliable way to rate the extent to which papers – or methods or policies – comply with the scientific method. This checklist, along with other checklists in this book, is also provided at GuidelinesForScience.com. To ensure that raters understood the guidelines, we pretested the checklist many times by examining the inter-rater reliability of the ratings for each of the criteria.

Checklist 3.1 is the result of our efforts: it provides 26 operational items to rate compliance with the eight criteria of the scientific method.

Checklist 3.1 *Compliance With Science Checklist*

Paper title:		

Reviewer:	**Date:**	**Time spent (minutes):**

Instructions for Raters
1. Skim the paper while you complete the checklist *as a skeptical reviewer.*
2. **Rate each lettered item,** below, marking the relevant checkbox to indicate
    True if the research complies,
    F/? (False/Unclear) if the research does *not* comply, or if you are unsure.
    IMPORTANT: If you are *not convinced* that the paper complied, rate the item **F/?**
3. If you rate an item **True,** *give reasons for your rating in your own words.*
4. **Rate criteria 1–8** as **True** by marking the checkbox only if all lettered items for the criterion are rated **T.**

First assess whether the paper complies with the lettered items under each criterion below. Then assess whether it complies with each of the eight criteria based on compliance with the lettered items. *Avoid speculation.*

1. Problem is important for decision-making, policy, or method development	☐ True T F/?
a. Importance of the problem clear from the title, abstract, result tables, or conclusions	☐ ☐
b. Findings add to cumulative scientific knowledge	☐ ☐
c. Uses of the findings are clear to you	☐ ☐
d. The findings can be used to improve people's lives without resorting to duress or deceit	☐ ☐
2. Prior knowledge was comprehensively reviewed and summarized	☐ True T F/?
a. The paper describes objective and comprehensive procedures used to search for prior useful scientific knowledge	☐ ☐

Checklist 3.1 cont'd

---

b. The paper describes how prior substantive findings were used to develop hypotheses (e.g. direction and magnitude of effects of causal variables) and research procedures	☐	☐

---

3. Disclosure is sufficiently comprehensive for understanding and replication	☐ True T	F/?

---

a. Methods are fully and clearly described so as to be understood by all relevant stakeholders, including potential users	☐	☐

---

b. Data are easily accessible using information provided in the paper	☐	☐

---

c. Sources of funding are described, or absence of external funding noted	☐	☐

---

4. Design is objective (*unbiased by advocacy*)	☐ True T	F/?

---

a. Prior hypotheses are clearly described (e.g., regarding directions and magnitudes of relationships, and effects of conditions)	☐	☐

---

b. All reasonable hypotheses are included in the design, including plausible naive, no-meaningful-difference, and current-practice hypotheses	☐	☐

---

c. Revisions to hypotheses are described, or absence of revisions noted	☐	☐

---

5. Data are valid (true measures) and reliable (repeatable measures)	☐ True T	F/?

---

a. Data were shown to be relevant to the problem	☐	☐

---

b. All relevant data were used, including the longest relevant time-series	☐	☐

---

c. Reliability of data was assessed	☐	☐

---

d. Other information needed for assessing the validity of the data is provided, such as adjustments, known shortcomings and potential biases	☐	☐

---

6. Methods were validated (proven fit for purpose) and simple	☐ True T	F/?

---

a. Methods were explained clearly and shown valid – unless well known to intended readers, users, and reviewers, and validity is obvious	☐	☐

Checklist 3.1 cont'd

	T	F/?
b. Methods were sufficiently simple for potential users to understand	□	□
c. Multiple validated methods were used	□	□
d. Methods used cumulative scientific knowledge explicitly	□	□
7. Experimental evidence was used to compare alternative hypotheses	**□ True**	
	T	**F/?**
a. Experimental evidence was used to compare hypotheses under explicit conditions	□	□
b. Predictive validity of hypotheses was tested using out-of-sample data	□	□
8. Conclusions follow logically from the evidence presented	**□ True**	
	T	**F/?**
a. Conclusions do not go beyond the evidence in the paper	□	□
b. Conclusions are not the product of confirmation bias	□	□
c. Conclusions do not reject a hypothesis by denying the antecedent	□	□
d. Conclusions do not support a hypothesis by affirming the consequent	□	□

Describe the most important scientific finding in your own words.

Sum the criteria (1–8) rated True for compliance: [ ] of 8

An electronic version of this checklist is available at guidelinesforscience.com.

3.2 Using the Checklist: For What, How, and by Whom

The *Compliance With Science Checklist* is intended to help researchers discover useful scientific knowledge and stakeholders to evaluate research.

As far as we are aware, the *Compliance With Science Checklist* is the only checklist designed for assessing the extent to which a paper complies with the scientific method. For example, a major US research funding body, the National Science Foundation, states that the agency was created by Congress in 1950 with a mission to "promote the progress of science" in its *Proposal and Award Policies and Procedures Guide* (National Science Foundation, 2019, p. viii), yet the agency does not define what it means by "science."

The *Compliance With Science Checklist* can be used for different purposes. For example, when it is used to assess whether to cite a paper for its scientific findings, researchers will find that they can typically complete the checklist in fewer than five minutes.

Researchers could also use the checklist to assess the compliance of their own papers before submission to a journal. Researchers should keep in mind that *they* are responsible for writing a paper that convinces raters that their research complied with the scientific method.

Before rating compliance with science for papers by others, raters should report potential biases, and sign an oath that: "I will rate this paper to the best of my ability and without bias." Raters who are uncomfortable signing such an oath, should not rate the paper.

An assessment of a research paper's compliance with all eight criteria takes less than half an hour on average. That estimate is based on the experiences of our research assistants, who rated more than 500 papers for us.

Anyone with a stake in useful scientific research can use the ratings from the completed *Compliance With Science Checklist*. Table 3.1 provides a list of potential users along with suggestions on how they could use the checklist.

For example, independent rating organizations could provide *Compliance with Science* ratings of papers as part of formal certification procedures, and for any of the other purposes listed in the table. Rating firms could meet the likely demand for ratings of university departments on the extent to which the research output of their researchers complies with the scientific method.

3.3   Not All Checklists Are Useful

If checklist items are irrelevant, misleading, or not based on scientific evidence or logic, the use of the checklist would be expected to harm decision-making.

Harmful checklists are often used in management. In one example, Porter (1980) proposed his "five forces" framework for competitive strategy planning based on opinions. To the best of our knowledge, the "forces" were not supported by experimental evidence, economic theory, or logic, as Rasmussen (2017) explained.

In another example, a series of laboratory experiments tested the value of the Boston Consulting Group's "BCG matrix," a four-item checklist for selecting investment opportunities. The subjects – 1,015 management students – worked independently. They were asked to choose between an investment opportunity that would double their investment and another that would lose half of their investment. Six researchers, each from a different country, ran experiments on 27 occasions during a five-year period. Of subjects exposed to the BCG checklist, 64 percent selected the unprofitable investment. Of those who were not exposed to the BCG matrix, 45 percent selected the unprofitable investment (Armstrong and Brodie, 1994). And, yes, it is a concern that only 55 percent of "unexposed" management students selected the profitable project.

3.4 Ensuring That Checklists Are Used

Consider the ARRIVE guidelines for animal studies. The 20-item checklist of guidelines was supported, but not required, by over 300 journals and major funders. A study of papers published in *PLOS* and *Nature* journals in the two years before and after the guidelines were introduced in 2010 suggested that "authors, referees, and editors generally are ignoring guidelines" (Baker et al., 2014, p. 1).

Is it sufficient to require completion of a checklist of guidelines? In a follow-up study on compliance with the ARRIVE guidelines, authors of 332 manuscripts were sent a copy of the checklist and told that they *must* complete it for their paper to be accepted, while authors of 340 manuscripts were not. There was little difference in the usage of the checklist and, despite the requirement, the papers of authors in the "must be completed" treatment were published regardless of whether they had done so. A completed checklist was requested again from the authors in the treatment group if they failed to comply the first time. Follow-up was not effective for compliance, either (Hair et al., 2018).

We find the lack of compliance with the requirement for completing a checklist strange. As experimenters, we have little trouble in getting subjects to complete checklists. We simply make it part of the contract. In our experimental studies, the subjects have always used the checklists as directed. In short, if a client states that payment for a project will only be made if a checklist is followed, we expect that nearly all who accept the contract and who are capable of completing the task will do so. We expect that if journals insisted on a completed checklist before considering a paper, they would have similar success.

4 ASSESSING THE QUALITY OF SCIENTIFIC PRACTICE

Gerd Gigerenzer, Director of the Max Planck Institute for Human Development, commented on reading papers in the *Journal of Experimental Psychology* from the 1920s and 1930s. He observed that "This was professionally a most depressing experience, but not because these articles were methodologically mediocre. On the contrary, many of them make today's research pale in comparison with their diversity of methods and statistics" (Gigerenzer, 2000, p. 296).

When Scott began his research in the 1960s, he easily found relevant and useful scientific papers by going to the university library, examining the shelves, and asking librarians. At that time, many faculty members who wrote papers did so because they wanted to. As a result, it seemed to Scott that the quality of papers was decent.

Nearly half century later, John Ioannidis – a leading health policy researcher – came to the conclusion that "most published findings are false" (2005a, p. 696). In his 2006 book – *The Production of Knowledge* – the long-time editor for the prestigious *Administrative Science Quarterly*, William Starbuck, came to the conclusion that "hundreds of thousands of talented researchers are producing little or nothing of lasting value" (2006, p. 3).

A 1990 survey asked editors of American Psychological Association (APA) journals: "To the best of your memory, during the last two years of your tenure as editor of an APA journal, did your journal publish one or more papers that were considered to be both controversial and empirical?" That is, papers that presented empirical evidence contradicting the prevailing wisdom. Sixteen of the

20 editors replied; seven could recall none, four said there was one, three said at least one, and two said they published several such papers. Over the 32 journal-years covered, only one paper with controversial findings received wholly favorable reviews. In that case, the editor revealed that he wanted to accept the paper, so he had selected reviewers who would be favorably disposed toward it (Armstrong and Hubbard, 1991).

In this chapter we describe how we went about reviewing the evidence on scientific practice. We discuss the evidence on whether journal articles provide information that helps readers to make better decisions. We also examine the topical, but always important, issue of the replicability of research studies and the related concerns of disclosure of methods and of data. We then examine evidence on the prevalence of cheating in scientific research and, finally, assess the trend in the rate of useful discoveries.

This chapter sets the groundwork for our more detailed examination of problems with research practice in Chapter 5 (Scientific Practice: Problem of Advocacy), Chapter 6 (Scientific Practice: Problem of Journal Reviews), and Chapter 7 (Scientific Practice: Problem of Government Involvement). Subsequent chapters describe solutions to the problems.

4.1   How We Reviewed Evidence on Research Practice

We reviewed surveys of research practices to gauge the use of the scientific method in papers published in academic journals. We searched the Internet and found relevant papers, followed up on their references, and then followed up on the additional papers that had useful findings, and so on – a technique known as snowballing. That approach was more successful at finding useful references than a one-time internet search (Armstrong and Pagell, 2003).

Reviews – especially meta-analyses – turned out to be especially useful. The earliest we found – Kupfersmid and Wonderly (1994) – covered many aspects of research in which practice fell short of the scientific method. Hubbard (2016) provided a review of about 900 papers, many related to what he referred to as corrupt scientific practices. Nosek and Bar-Anan (2012), Nosek et al. (2012), and Munafo et al. (2017) provided reviews that together cited hundreds of publications on the practice of science. Those sources also suggested ways to correct the deficiencies. We include many of their suggestions in our chapters of solutions, starting with Chapter 8 (What It Takes to Be a Good Scientist).

To ensure that our descriptions of other researchers' work were accurate, we emailed the authors of each paper whose substantive findings we cited. If we received no reply, we sent a reminder. Our messages included our descriptions of their work. We asked whether we correctly summarized their findings and, if not, how we could improve our summary. We also emphasized that we were interested to learn of scientific findings that might challenge our conclusions.

We received replies from 76 percent of the authors we were able to contact. Roughly a third of those who replied also provided suggestions for improvements to our descriptions or provided references for additional evidence.

4.2   Do Journal Papers Help Readers to Make Better Decisions?

The usefulness of research papers published in scientific journals has been a cause of concern for decades. A 1991 study examined whether those familiar with scientific research on consumer behavior were better able to make predictions of consumer behavior than those who were not familiar. Predictions were obtained for the outcomes of 105 hypothesis tests from 20 empirical studies published in the *Journal of Consumer Research*. Academics (16), practitioners (12), and high school students (43) made 1,736 predictions in all. Of these, 51.3 percent of academics', 58.2 percent of practitioners', and 56.6 percent of students' predictions were correct. A survey of the 43 members of the *Journal of Consumer Research*'s editorial board was conducted to assess their expectation on the results of the study. They expected correct predictions to be made by academics for 80 percent of hypotheses, by practitioners for 65 percent, and by high-school students for 55 percent (Armstrong, 1991).

An earlier study surveyed editors of psychology and social work journals, asking them to rank criteria for selecting which papers to publish. Usefulness – "the value of an article's findings to affairs of everyday social life" – ranked 10th in importance out of 12 criteria (Lindsey, 1978, pp. 18–21).

The relative lack of interest in usefulness among scientific journal editors is not confined to the social sciences. A mail survey of the editors of the top ten journals in physics and chemistry, as well as sociology and political science, found that the "applicability to practical or applied problems" ranked last of the ten criteria presented for all but political science, for which it was next to last (Beyer, 1978).

As far as we are aware, there has been no shift in journal policies toward publishing papers that are useful, in the sense that they can help readers to make better decisions. Moreover, the number of journals and of articles published has increased dramatically, presumably making it more difficult to find papers that provide findings that are useful to decision makers.

4.3   More Replications of Important Papers Needed

Replications are needed to establish findings as valid. The requirement applies no less to the social sciences, as Schmidt (2009) maintained: "To confirm results or hypotheses by a repetition procedure is at the basis of any scientific conception ... proof that the experiment reflects knowledge that can be separated from the specific circumstances ... under which it was gained" (p. 90). They are also needed in order to obtain sufficient observations from which to derive accurate estimates of effect sizes using meta-analyses.

To assess the relative frequency with which replication and extension papers are published, Hubbard and Vetter (1996) sampled 472 issues drawn from 18 leading journals in accounting, economics, finance, and management published between 1970 and 1991. Their initial sample was 6,400 research papers, of which 4,270 were empirical. They found that only 6.2 percent (266) of the empirical papers were replications with extensions.

A follow-up to a study of replications in three major marketing journals (Hubbard and Armstrong, 1994) found that the proportion of empirical studies that could be counted as replications, with extensions, declined from 2.4 percent in the period 1974 to 1989 to 1.2 percent in the period 1990 to 2004 (Evanschitzky et al., 2007).

In spite of the scientific requirement for replications of important papers, journal reviewers are biased against them on the basis that they offer nothing new if the original findings are supported and that the replicators must have done something wrong if they are not. As a consequence, researchers, who need publications in high status journals to meet performance criteria, are discouraged from conducting them (Evanschitzky and Armstrong, 2013).

4.4   Are Papers in Scientific Journals Replicable?

The ability to replicate a study is,
typically, the gold standard by which
the reliability of scientific claims is judged.
     US National Research Council
     (2002, p. 7)

There is a caveat to the above quotation. That is, judgments on reliability apply only to papers that use the scientific method. Advocacy can always provide "successful" replication.

Replications and extensions are critical for scientific findings. High-status scientific journals invariably state that the papers they publish must be replicable. Without the *possibility of replication, no paper can claim that its findings are scientific.*

Replications are only useful for *important papers*. For example, Milgram's obedience to authority studies were so important that many researchers conducted replications (Blass, 2009). Some extensions helped find ways to reduce blind obedience to authority. One extension used role-playing and found that irresponsible behavior by corporations was reduced when different stakeholders were represented on the board of directors, and when the corporation's accounting system included the costs and benefits for each stakeholder group (Armstrong, 1977). Another extension found that when cues present in Milgram's original experiment briefing of subjects that the "learner" was not actually being seriously hurt were removed, none of the 10 subjects was fully obedient (variation DM-8 in Mixon, 1972).

Direct replications are vital for papers that challenge prior knowledge on important problems, or where fraud is suspected. They are not useful, however, for papers that do not follow the scientific method.

Extensions, or replications with variations, can be used to test the limits of findings. They help to assess the generalizability of the findings by testing the same phenomena under different conditions. A failure to replicate can be useful for judging the conditions under which the original findings apply.

Failures to replicate can be due to errors in the original study or in the replication. They could also be due to unknown problems in the measurement instruments. McCullough (2000) found different results for the same computation when using different statistical software. Boylan et al. (2015) found similar but not exactly the same results in a replication of a forecasting study either due to a small sample size, or to communication problems in understanding the procedures used in the original study.

4.4.1 Replications Can Reverse Influential Prior Findings

Hirschman's (1967) influential "hiding hand" study of 11 public works projects financed by the World Bank found that while planners underestimated costs, they underestimated benefits even more, suggesting that public-works projects are typically beneficial. Half a century later, Flyvbjerg (2016) replicated Hirschman's research by analyzing 2,062 projects involving eight types of infrastructure in 104 countries during the period from 1927 to 2013 and found that there was no "hiding hand" benefit. On average, costs overran by 39 percent and benefits were overestimated by 10 percent.

In a paper with more than 2,000 citations on Google Scholar, Staw (1976) claimed that managers are subject to escalation bias when making investment decisions because they continue to invest in projects that are performing poorly. In other words, they throw good money after bad. A replication failed to reproduce the findings (Armstrong, 1996; Armstrong et al., 1993). The replication has been little cited; roughly 50 times on Google Scholar. Meanwhile, escalation bias continues to be taught as a judgmental failure in some universities, despite the fact that a rational solution, known for decades, is to calculate the net present value of alternative investments when a decision is needed.

A highly cited study on buyer behavior showed that when shoppers were offered 24 jams, fewer than 3 percent made a purchase, whereas when they were offered only six jams 30 percent purchased. The researchers concluded that customers should not be offered too many choices (Iyengar and Lepper, 2000).

An attempt at direct replication of the jam choice study failed (Scheibehenne et al., 2010). The authors of the replication attempt also conducted a meta-analysis of 50 related empirical studies and found an effect size of zero, indicating that there is no evidence for the "too-many-choices" effect.

Further studies have shown that the number of choices that consumers prefer is affected by many factors (Armstrong, 2010, pp. 35–39). With many choices available shoppers need more time to shop and must expend effort in making decisions. To deal with that problem, Trader Joe's invites suppliers to compete for shelf space on the basis of winner takes all, thus providing the products buyers prefer, and saving on floor space and shoppers' time.

Failures of extensions are often useful, as they can help to identify conditions that affect the relationships. For example, experimental studies on humor in advertising have identified some conditions under which humor is persuasive, and other conditions under which humor reduces persuasiveness (Armstrong, 2010).

The reversal of influential earlier findings is not confined to the social sciences. An analysis comparing the findings of subsequent studies with 45 highly-cited original clinical research studies found that 16 percent were contradicted by the later studies. Another 16 percent had reported stronger effects than were found by the later studies (Ioannidis, 2005b).

4.4.2   Criteria for a Successful Replication

One modest yet universal criterion for a successful replication is that the findings agree on the directions of the effects. Another reasonable criterion is that the effect sizes found in replications should resemble those in the original study. As we discuss later in this book in Chapter 6, statistical significance tests should never be used to judge replication success, nor for any other purpose.

For direct replications, where the same procedures are used, the findings should be very similar. Extensions, on the other hand, are helpful in assessing "construct validity," or the degree to which a test measures what it claims to be measuring. Do the same findings occur for experiments in varied conditions? Also important is that failed extensions might help to identify conditions that were previously overlooked.

4.4.3   Conditions Favoring Successful Replications

Papers complying with the criteria for the scientific method should have high replicability. In particular, replication success will tend to be high for properly designed experimental studies. That applies to all types of experimental studies: laboratory, field, natural experiments, and, to a lesser extent, quasi-experimental studies – i.e., experiments in which some, but not all, important causal variables can be controlled.

To test the validity of quasi-experiments, Armstrong and Patnaik (2009) analyzed 56 persuasion principles by using data from 240 pairs of print advertisements. The data controlled for target market, product, size of the advertisement, media and, in half the cases, brand. The advertisements differed, however, with respect to illustrations, headlines, and text. The sample sizes of the quasi-experimental studies were small, ranging from 6 to 118 tests with an average of 31 tests.

The directions of causal effects from the quasi-experimental analyses were consistent with those from field experiments for all seven persuasion principles for which comparisons were available. The directions were also consistent for all 26 principles from laboratory experiments. Finally, they were consistent with all seven principles supported by both field and laboratory experiments. In short, quasi-experimental findings always agreed with experimental findings on the direction of the causal effects. In contrast, the directional findings from non-experimental analyses were consistent with experimental data analyses for only 67 percent of the comparisons.

Another study found that research by economists using non-experimental data suggested that top corporate executives of publicly listed firms are not overpaid. However, experimental studies by researchers in organizational behavior found that CEOs are, in fact, overpaid and that the excessive pay is detrimental to the performance of the firms and thereby to their stockholders (Jacquart and Armstrong, 2013).

Kabat's (2008) book on environmental hazards – including the insecticide DDT, electromagnetic fields from power lines, radon, and second-hand smoke – concluded that analysis of non-experimental data in studies – along with a failure to consider alternative hypotheses – has led to false conclusions about their effects, and thereby misled researchers, doctors, patients, producers, and the public.

4.4.4   Failures of Disclosure Undermine Replication

Without full disclosure of methods and data, direct replication is impossible. How, then, do researchers respond to requests for data and materials needed to replicate their research?

To answer that question, the authors of one study sent requests that appeared to be from a graduate student to a sample of senior researchers who had published in 1978 or 1979 "data-based articles" in one of five journals in the field of marketing. The final sample for analysis was 99 requests. One half (49) responded to say that the materials were available, 14 percent responded that they were not, and 36 percent did not reply (Reid et al., 1982).

In an attempt to replicate empirical economic research papers that had been published in the *Journal of Money, Credit and Banking*, Dewald, Thursby, and Anderson (1986) sought computer programs and data from the authors of the papers. They received responses from 42 of 62 authors. Among the responding authors, nearly half (48 percent) failed to supply data, with 5 percent claiming confidentiality, 33 percent stated that they were lost or destroyed, and 10 percent did not provide data but claimed they were available from published sources. In all, they received data from the authors of only 35 percent of the published papers for which they had requested the data. There were problems with all but one of the data sets they did receive; errors in transcription, data transformation, and computer programming were commonplace, and some of the errors substantively affected the conclusions of the studies examined. The authors of the study concluded that none of the replications reproduced the original results with good precision.

In 2015, the Open Science Collaboration (OSC) published a study that asked 270 contributing authors to replicate 100 studies published in three reputable psychology journals in 2008 (*Psychological Science, Journal of Personality and Social Psychology*, and *Journal of Experimental Psychology: Learning, Memory and Cognition*). The studies were chosen for timeliness, ease of replicability, and representation of various fields within psychology (Open Science Collaboration, 2015).

Media reports and OSC's own summary claimed that fewer than 40 percent of the studies were successfully replicated. That finding was roughly in line with Hubbard's (2016, pp. 141–142) meta-analysis of 804 replication outcomes in 16 studies over seven areas of management science. He found that 46 percent of the replications conflicted with the original papers.

The journal *Nature* followed with a survey of 1,576 researchers. About half agreed that there was a "significant crisis" with replications (Baker, 2016). In contrast, the authors of the OSC replications judged 84 percent of the findings and 71 percent of the effect size estimates to be at least somewhat similar to those of the original studies.

Why such a difference in opinions? We believe that it has to do with the misconception that statistical significance is a measure of reproducibility. It is not. That belief, common even among researchers and teachers of statistics, is known as the replication delusion, or replication fallacy (Gigerenzer, 2018a).

Using OSC data, we determined whether each replication agreed with the original study's effect direction. After simply calculating which percentage of the studies had the same direction and which had the opposite, we were able to determine the success rate of those replications on that basis. In total, 82 percent of the replications found the same directions of effects as the original study, which is a close match with the judgments of those who did the replications.

4.5   How Much of a Concern Is Cheating?

Scientists have long been known to cheat; even famous scientists including Cyril Burt, Galileo, Mendel, and Newton have been suspected (Broad and Wade, 1982; Westfall, 1973). The frequency of cheating, however, was apparently low – for example, a 1982 survey of 13 management science journal editors found them to be generally unconcerned about cheating (Armstrong, 1983). After all, cheating damages a scientist's reputation if it becomes known.

Cheating by scientists was recently examined in the book by Chevassus-Au-Louis, *Fraud in the Lab* (2019). He attributed cheating to the incentives faced by scientists, including the need to publish and the expectation that research findings would be consistent with the preferences of the funders.

The rate of journal retractions, an indication of cheating, was around 1 in 10,000 in medical research from 1970 to 2000. It grew by a factor of 20 from 2000 to 2011 (Brembs et al., 2013). There are a number of possible explanations. One is that the Internet makes it easier to detect cheating. Another is that more journal articles are being published.

An analysis of 2,047 biomedical and life science articles published after 1972 and retracted by May 3, 2012, revealed that 67 percent of the retractions were due to misconduct. The rate of retractions due to misconduct increased roughly tenfold between 1975 and May 2012 (Fang et al., 2012).

Are faculty members aware of cheating? Bedeian et al. (2010) conducted an internet survey of 1,940 tenured and tenure-track management faculty associated with 104 PhD granting management departments of accredited US business schools. They asked whether the respondents had "observed or heard about" faculty engaged in "fabrication, falsification and plagiarism" and "questionable research practices" such as duplicate publication of data. They analyzed 384 usable responses. Some of their key results were: (1) "withheld methodological details or results," 79 percent; (2) "selected only those data that support a hypothesis and withheld the rest," 78 percent; and (3) "withheld data that contradicted their previous research," 50 percent. They also found that 92 percent claimed to have observed or heard about researchers who, within the previous year, had developed hypotheses *after* a study's results were known. Admittedly, those findings only state that many researchers believed those practices had occurred, and the authors caution that "to the extent that more than one respondent may have reported on the same acts" their results do not necessarily document the actual frequency of the reported behaviors.

Another survey of over 2,000 psychologists found 35 percent admitted to "reporting an unexpected finding as having been predicted from the start." Further, 43 percent had decided "whether to exclude data after looking at the impact of doing so on the results" (John et al., 2012, p. 525, tab. 1).

Papers in high-status journals have been found to be *more likely* to include fraudulent research than those in lower-ranked journals (Fang et al., 2012). Might that occur because researchers obtain greater status, grants, salary, and promotions if they publish in such journals?

A meta-analysis of 18 surveys of scientists found that about 2 percent admitted to misconduct that is harmful to *science* – the figure did not include plagiarism, which primarily harms *scientists*. In addition, 34 percent admitted to "questionable research practices" such as hypothesizing after the results were known, or HARKing. Another approach, "*p*-hacking," involves trying different combinations of variables, functional forms, removal of outliers, or testing different time periods – also known as "cherry picking" – to get the desired results (Fanelli, 2009).

Another strategy is fabrication: John Darsee – a medical researcher at Emory and Harvard – admitted to fabricating data for a published paper. An investigation concluded that he had fabricated data in 109 publications involving 47 other researchers. Some of the fabrications were preposterous, such as a paper using data on a 17-year-old father who had four children, ages 8, 7, 5, and 4. The papers were published in leading peer-reviewed journals (Stewart and Feder, 1987).

Data fabrication is a long-standing malpractice, with the aforementioned Cyril Burt's (1883–1971) suspected fabrication of twins' IQ data in order to support his eugenics hypotheses an infamous example. Babbage described earlier incidents (Chevassus-Au-Louis, 2019). In the cases that have been uncovered, the frauds typically led first to great success, which was later followed by disgrace. Creating or manipulating data is, in the long run, a poor strategy for anyone who aspires to be a scientist.

In a hoax, a computer program called SCIgen randomly selected complex words commonly used in a topic area, and then used grammar rules to produce sentences and paragraphs. The software-generated text was first used to test whether reviewers would accept meaningless papers for a conference. The title of one such paper was "Simulating Flip-flop Gates Using Peer-to-peer Methodologies." Some of the software-generated papers were accepted. Later, researchers, seeking to pad their résumés, used the SCIgen program to submit papers to scientific journals. In the order of 120 SCIgen generated papers were published in respected peer-reviewed scientific journals before they were discovered and withdrawn (Lott, 2014).

We describe solutions for, or at least ways to reduce, cheating in Chapter 9 (How Scientists Can Discover Useful Knowledge) under the heading "Selecting a Problem" in a subsection on replications, in Chapter 11 (How Stakeholders Can Help Science) under the subheadings "Employ an Ombudsman," "Require an Oath That Standards Were Upheld," and "Use Benford's Law to Identify Data Fabrication." Note that all but the last of the solutions require the full disclosure of methods and data.

4.6   How Efficient Is Research?

Not very. And research has been getting increasingly inefficient.

Holub et al. (1991) suggested an iron law of research in economics: "the number of important articles in a field of economic theory increases by the square root of the total of all articles published in this field" (p. 317). They defined important papers as those that had been cited at least 30 times. Their data were 2,681 articles on growth theory that had been published between 1939 and 1986 in 46 journals important to economists. As many as 44 percent of the articles had not been cited at all, and only 2 percent were important according to their definition. They estimated the relationship between the number of important papers published and the total number of papers published using that data.

Armstrong and Pagell (2003) examined progress in research on forecasting methods over the latter half of the twentieth century. They concluded that only 2.5 percent of papers over that period provided useful findings, a figure similar to the estimate of Holub et al.'s (1991) above.

5 SCIENTIFIC PRACTICE: PROBLEM OF ADVOCACY

Objectivity underlies all of the criteria for complying with the scientific method. Only an objective description of an objective study and its findings can be considered scientific. Yet objectivity does not come naturally, and so scientists must struggle to overcome a tendency to advocate for subjectively preferred hypotheses when they practice science.

In this chapter, we discuss the unnaturalness of objectivity and the tendency to confirm rather than to test initial hypotheses. We then describe ways in which the opposite of objectivity, advocacy, is practiced, and the extent to which it is practiced and accepted.

We provide solutions for advocacy in Chapter 9 (How Scientists Can Discover Useful Knowledge), Chapter 10 (How Scientists Can Disseminate Useful Findings), and Chapter 11 (How Stakeholders Can Help Science). Key among the solutions are the practices of testing multiple reasonable hypotheses, and of providing full disclosure so as to facilitate checking and replication attempts.

5.1 Unnaturalness of Objectivity

It is easier for people to believe lies than
to convince them that they have been lied to.
Mark Twain

People have preconceptions and expectations, and they prefer to avoid information that might contradict those beliefs. Those human traits strain our ability to objectively assess new evidence.

Imagine being asked to conduct a scientific study on the value of education. Could you be objective and design a study that *might* conclude that a formal education is likely to be harmful to most students? If the results indicated that the treatment was harmful, would you believe it? Would you publish the study?

In 1939, a sample of 506 boys with a median age of 10.5 years was assigned either to a five-year program that included tutoring in academic subjects and individual counseling twice a month, or to a control group that was only monitored during their time at school. This "Cambridge-Somerville Youth Study" was a well-designed experiment: each subject was matched with the boy who was most similar to him. The treatment group was randomly selected by choosing one boy from each pair. The study then tested whether providing additional support beyond normal schooling would help the boys in the treatment group.

A follow-up study by McCord (1978) located 208 members of the treatment group and 202 members of the control group. Questionnaires were completed by 113 of those in the treatment group, and by 122 in the control group. The responses revealed that those in the treatment group were positive about the value of the program, with two-thirds reporting that it had been helpful for them.

However, their opinions were not consistent with the outcomes of the program. Those in the treatment program were more likely to have committed a serious crime, been an alcoholic, been employed in low-status occupations, reported low job satisfaction, or died young. There were no measured outcomes showing improvements

McCord concluded that the treatment group suffered relatively poor outcomes because they had learned to depend on others, rather than to take responsibility for their own learning. Her conclusions upset researchers and policy makers. When she presented her findings, professors in one lecture yelled at her.

Her conclusion challenges the major assumption in schools. Universities routinely ask students to rate their professors on their "teaching," including questions about their ability to motivate students to learn. Research since McCord's paper has continued to challenge the benefits of education to students when someone else is responsible for their learning (see Armstrong, 1980d). Moreover, reviews of evidence have challenged the assumption that a university education contributes to the economic growth of nations (e.g., Armstrong, 2012c; Wolf, 2002).

People tend to avoid information that is inconsistent with their beliefs. If they cannot avoid new information that challenges their beliefs, they tend to evaluate it in ways that leave their prior beliefs unchanged. If they have no prior beliefs, they tend to adopt the beliefs of their friends, authorities, or groups with whom they identify.

Experimental studies have shown that groups typically reject people with opinions that differ from the beliefs of people in the group. That happens even for newly formed groups when asked to decide on something about which they had no prior opinion. Pressure is directed against deviants, who face ostracism if they do not quickly agree with the group.

Subjects in an experiment were put in groups to discuss what should be done with a juvenile delinquent, Johnny Rocco. Possible actions ranged from giving Johnny a more nourishing, loving environment, to sentencing him to severe disciplinary action. A confederate of the researcher avoided offering an opinion until it was clear that the rest of the group reached a consensus. Then he dissented. The group tried to change his opinion. When that did not work, they ostracized him. And when the group nominated members for committee positions, they overwhelmingly nominated the dissenter for the least desirable and least important role (Schachter, 1951).

In another experiment, members of a Christian youth group – most of whom were strong believers that Jesus was the Son of God – were presented evidence, ostensibly from the Dead Sea Scrolls, that Jesus was not divine. Among the group, those who believed the disconfirming evidence to be authentic reported the biggest *increase* in their belief in the divinity of Jesus (Batson, 1975).

For an example of the persistence of an extreme belief, researchers studied a cult that predicted the "end of world" on a particular date. When that did not happen, the cult members *increased their belief* that the world would end, and set a new date (Festinger et al., 1956).

Human behavior does not seem to change over time. MacKay (1841) used 740 pages to describe "Extraordinary Popular Delusions."

5.2   Confirmation Bias

> ... when any proposition has been once laid down
> ... [it] forces everything else to add
> fresh support and confirmation.
>> Francis Bacon
>> (1620, Aphorism XLVI)

Confirmation bias is the tendency to interpret new evidence to support one's existing beliefs. It also involves efforts to seek confirming evidence and to avoid challenging information. If a challenge cannot be avoided, the source of the challenging information is dismissed as untrustworthy.

To assess confirmation bias, 29 undergraduate psychology students were provided with three numbers (2, 4, 6) and were asked to identify the rule used to generate that sequence. The subjects could obtain information by conducting experiments – proposing their own sequence of three numbers – in order to discover the rule. After each experiment, they were told whether they had correctly identified the rule. Most subjects persisted in searching for evidence to support their initial hypothesis, to the neglect of testing alternatives (Wason, 1960).

Other researchers using Wason's "2-4-6" problem found that subjects continued to have confidence in a hypothesis even when more than half of their predictions were wrong. The subjects tended to believe the feedback when it was positive but ignored it when it did not confirm their hypotheses (Mahoney and DeMonbreun, 1977).

The 2-4-6 problem was also used in experiments with scientists as the subjects. Many psychologists and physical scientists proposed number sequences that they expected would confirm their hypotheses. Fewer than 40 percent of the scientist subjects tested their hypotheses by seeking disconfirming evidence. Overall, only 10 percent of the subjects sought disconfirming evidence (Mahoney and DeMonbreun, 1977).

Another 2-4-6 experiment contrasted the ability of 15 psychologists and 15 physical scientists with that of 15 protestant ministers. The scientists were quicker to develop hypotheses. They were also more tenacious in sticking with their hypothesis after it had been shown to be false (93 percent) than were the ministers (53 percent) (Mahoney and DeMonbreun, 1977).

Wason's finding that most subjects fell victim to confirmation bias raises further concern about the obstacle that presents to the practice science. There are, however, caveats to the concern. Wason's problems were context free abstractions. The subjects were given no information that would help them to relate the problems to real, familiar problems. There was no information on the absolute and relative costs and benefits of obtaining information. Nor was there information on the subject's role.

Experiments that tested the effects of adding content and context to the Wason selection task found that most subjects made choices that were logical given their role and the situation that was described to them. For example, 83 percent of subjects chose to turn over cards corresponding to "benefit taken" and "cost not paid" when they were asked to be in the role of an interested party in a social contract situation (Ortmann and Gigerenzer, 1997). Thus, it would appear that the effect size is – for practical situations absent perverse incentives to confirm hypotheses such as are common in academia – likely substantially smaller than originally estimated.

5.3   How Advocacy Is Practiced

Advocacy is often easy to identify. For example, Scott was asked by a US Senate Committee to testify on a report with the title, *USGS Science Strategy to Support U.S. Fish and Wildlife Service Polar Bear Listing Decision*. The listing decision was not supported by an audit of the methods that were used to forecast the polar bear population (Armstrong et al., 2008).

Advocates use various tactics to support the hypotheses that they or their client desires. In discussing these tactics, we use experiments by Kahneman and Tversky (K&T); in particular, their 1979 "Prospect Theory" paper, which is the most cited paper in economics and the third most cited paper in psychology (Simonsohn, 2014). The theory states that people tend to be more concerned about losses than gains, such as that losses of US$1,000 would, on average be for people as important to them as gains of US$2,000. This would appear to lead to irrational decision-making. For their body of research, Kahneman was awarded the Nobel Prize in Economic Sciences in 2002.

One might expect Nobel Prize-winning research to exemplify good scientific practice. Surprisingly then, when we employed five research assistants to rate the prospect theory paper for compliance with science using Checklist 1, the paper rated as compliant with only the first of the eight criteria: importance of the problem (Table 5.1).

Table 5.1. *Prospect theory's* compliance with science (consensus of five novice raters)

Criterion	Rater agreement (%)	Compliant with science
1. Important problem	80	Yes
2. Comprehensive literature review	80	No
3. Sufficient disclosure	60	No
4. Objective design	100	No
5. Valid and reliable data	80	No
6. Valid simple methods	60	No
7. Experimental evidence	80	No
8. Logical conclusions	60	No

One of Scott's earliest recollections about the biases and heuristics research that K&T pioneered was a colleague who said to Scott that, "In the past when we developed questionnaires, much time was spent for revising questionnaires to ensure that respondents understand the questions in surveys. Now, in the biases and heuristics literature, if the subjects understand the question, the researchers revise the questions to get the answers they are looking for."

Christensen-Szalanski and Beach (1984) suggested that the study of biases in judgment and decision-making is itself biased. In particular, they pointed out that while findings on performance in judgmental tasks include examples of both good and bad performance, popular summaries of the research and introductions to journal articles leave readers with the impression that the research findings "clearly show human judgment and decision making to be hopelessly inadequate" (p. 75). To test their hypothesis, they coded more than 3,500 abstracts of empirical studies related to good or bad judgments or decisions taken from *Psychological Abstracts* from 1972 through 1981. On average, studies finding poor judgment were cited almost six times more than those that found good judgment.

They then surveyed judgment and decision-making researchers, asking them to recall examples of studies that found people's judgments were good and others that found that they were bad. The few studies identified that found judgments were poor were all laboratory studies, mostly with college students as subjects. In contrast, 58 percent of the studies identified as finding good judgment were conducted in applied settings such as in weather forecasting.

Gigerenzer (1991) noted that the judgment problems used in K&T's biases and heuristics research differed from people's everyday experiences of such problems. He tested what would happen if the questions were posed in more familiar and realistic ways. When he did so, he found the subjects made judgments and decisions that were rational given the contexts of the situations that were described to them. He concluded that K&Ts framing of problems in ways that subjects tended to interpret as decisions under uncertainty – as opposed to decisions with known possible outcomes and risks – was responsible for the subjects' responses, rather than failings in human reasoning (Gigerenzer, 2018b).

Similarly, Vernon Smith – who shared the 2002 Nobel Prize with Kahneman – concluded that research in experimental economics had found that "most standard theory provides a correct first approximation in predicting motivated behavior in laboratory experimental markets ... but is not plainly represented in contemporary research in economic psychology (Smith 1991, p. 887).

In regard to prospect theory in particular, a meta-analysis of 136 papers with experiments involving in the order of 30,000 subjects found that it was difficult to determine the conditions under which prospect theory applied (Kühberger, 1998). More recently, Locke and Latham (2019) examined experimental evidence on the claim that prospect theory provides a useful explanation of goal directed behavior. They found that the claims for prospect theory are based on judgments about ambiguous and unrealistic scenarios, and that the theory fails to describe the effect of goals on motivation and outcomes in real human situations.

5.3.1 Showing Only Evidence Favoring Hypothesis

One study reviewed the literature cited by those urging governments to "nudge" citizens to behave in preferred ways – such as by requiring people to actively opt out of a government-chosen "default." The study found that papers advocating governmental nudges misinterpreted the effects of flawed experimental design on subjects' decisions and mistook logical norms for systematic failures in human reasoning. They ignored evidence that normal human judgments are not systematically biased, and that people do not need to be "nudged" by the government to behave in ways that are consistent with their welfare (Gigerenzer, 2015).

Another example relates to the effect of eating red and processed meats on health outcomes. In an editorial titled "Meat consumption and health: food for thought" announcing the findings of meta analyses, the authors noted of epidemiological studies in the field that many "selectively report results" (Carroll and Doherty 2019, p. 767). The overall conclusion from the meta-analyses introduced by the editorial was that there was a lack of evidence of sufficiently high quality to support claims that eating red and processed meats has adverse health effects.

The meta-analyses were published against the objections of other researchers in the field who had been advocating for reduced red meat consumption. They urged the journal editor not to publish the papers. When she would not cooperate, they complained to the Federal Trade Commission, and then to the Philadelphia district attorney's office (Laframboise 2020a, 2020b).

When there is little or no objective evidence on a problem, an observation from a small study or even an opinion can, when published, be misconstrued and cited as support for a favored hypothesis. More citations of the original publication and citations of the citing works create a fiction of evidence for the hypothesis.

Does that happen in practice? Often enough to be named the Woozle effect after an incident in A. A. Milne's *Winnie-the-Pooh* book in which two characters follow the ever-increasing tracks of the Woozle ... which were in fact their own footprints in the snow as they went round and round a tree. A Google Scholar search found 123 results in February 2020. The phenomenon is also known as evidence by citation. *Wikipedia* describes the origins and provides examples.

5.3.2 Ignoring Cumulative Scientific Knowledge

A problem related to the previous one is where researchers frame their problem as unique, new, or in some way importantly different from what has gone before or previously been studied. Take the law of demand, for example.

A basic law of economics is that people are less willing to purchase goods and services at higher prices than at lower prices, all else being equal. In short, common goods have negative price elasticities of demand. A meta-analysis of price elasticities of demand[1] for 1,851 goods and services from 81 studies found that the average elasticities ranged from –0.25 to –9.5, with half of the studies' estimates between –1.80 and –3.26, and an overall average of –2.62 (Bijmolt et al., 2005, tab 1).

Contrast that with findings of an average elasticity estimate of just –0.2 from a review of 1,474 studies of the price elasticity of demand for common unskilled labor services (Doucouliagos and Stanley, 2009, p. 412). How can one explain the implication that employers barely respond to changes in the cost of labor?

5.3.3   Avoiding Specifying Conditions That Admit Rejection

A hypothesis that is specified with neither conditions nor a statement that it is unconditional is unscientific as it cannot be falsified yet may never be true. As we described above, a meta-analysis of 136 papers with experiments involving 30,000 subjects found that it was difficult to determine the conditions under which prospect theory applied (Kühberger, 1998).

Another example is Aristotle's proposal that a two-sided argument is more persuasive than a one-sided argument. Later experimental research found that the proposed rule did not apply when opposing arguments were raised *but not refuted* (Allen, 1991; O'Keefe, 1999), nor did it apply when people were already favorably disposed toward the message, or when the opposing argument was presented first (O'Keefe, 1999).

5.3.4   Avoiding Tests of ex ante Predictive Validity

Consider again K&T's "prospect theory." We were unable to find any attempts by K&T to test the ex-ante predictive validity of prospect theory applied to realistic problems. Still, Kahneman was

---

[1] A price elasticity of demand for a good or service can be thought of as the percentage change in the quantity demanded that results from a 1 percent change in the price. They are expected to be negative (higher price causes a lower quantity to be demanded) and larger (more negative) in the longer term as buyers change their plans and find substitutes, and larger if the cost of the good or service is a large proportion of the buyer's budget.

confident of their findings. For example, Kahneman (2011, p. 300) stated that, "The concept of loss aversion is certainly the most significant contribution of psychology to behavioral economics." However, other researchers had tested the out-of-sample predictive validity of prospect theory:

- A review of four experiments on healthcare messages framed as either losses or gains found that prospect theory was unhelpful for developing persuasive messages (Wilson et al., 1988).
- As described above, a review of 136 papers with about 30,000 subjects found that it was difficult to determine the conditions under which prospect theory would work (Kühberger, 1998).
- The authors of a review of studies on the effect of positive versus negative framing on judgments and decisions concurred with Kühberger that "the variety of framing phenomena cannot be understood adequately within ... prospect theory" (Levin et al., 1998, p. 179). Moreover, they found that 28 studies concerned with the "risky choice framing" type of problem that they hypothesized would provide the best fit with prospect theory were "challenging to the development of a uniform theoretical account" (p. 155).
- A meta-analysis of experiments with over 50,000 subjects and 165 effect size estimates found that predictions derived from prospect theory were not confirmed (O'Keefe and Jensen, 2006).

K&T did not inform researchers about predictive validity failures. For example, in Kahneman's 2011 book, *Thinking, Fast and Slow*, we could find no mention of the reviews described above. That oversight violates criterion 2 in the *Compliance with Science Checklist*, which requires prior knowledge to be comprehensively reviewed and summarized.

K&T are not unique in this. For example, in the field of health and cancer risk assessment, Hermann Muller suppressed knowledge that his Linear-No-Threshold hypothesis on the relationship between radiation and cancer was inconsistent with the best experimental evidence available at the time even as he accepted his Nobel Prize (Calabrese, 2019).

5.3.5   Ignoring Important Causal Variables

Schmidt (2017) showed that desired policies are often supported by ignoring the importance of key variables. For example, the extent to which the higher earnings of college graduates compared to high-school graduates can be explained by differences in general mental ability. Ignoring that evidence leads to the erroneous conclusion that schooling is a major determinant of future earnings.

5.3.6 Failing to Use Experimental Evidence

In 1974, Scott was invited to visit the University of Illinois to interview for a chaired professorship. At the time, Julian Simon was a professor in that department. He was Scott's role model and they had kept in touch often yet had not previously met. When Scott arrived at his office, Julian said "Don't sit down. We are going out for a walk." So out they went and stopped at clearly laid out plots of land. Julian explained that they were used to experiment with various agriculture procedures. "That is why agriculture continues to make such rapid scientific advances every year." Julian paused. Scott thought about it for less than about a minute and then said, "understood." Much of Scott's previous work had involved non-experimental data. He resolved to study problems using experimental data whenever possible.

While analyses of non-experimental data might, like the judgments of domain experts, produce useful ideas, they cannot, on their own, produce scientific findings.

Why not rely on *non*-experimental data? Let's say that you asked a research assistant to conduct an experiment. When the assistant reports back to you, you discover that an important causal variable was neglected, some causal variables varied together rather than independently, and one of the variables was spurious. Could the collected data still be analyzed in good faith? Proceeding in that situation would be analogous to drawing conclusions about causality from an analysis of non-experimental data.

5.3.7 Using Machine Learning Methods

Machine learning methods, such as stepwise regression, neural networks, and data mining, are used to select "predictor variables" based on correlations in non-experimental "big data." Einhorn (1972) likened such data models to alchemy, stating, "Access to powerful new computers has encouraged routine use of highly complex analytic techniques, often in the absence of any theory, hypotheses, or model to guide the researcher's expectations of results" (p. 367.) More recently, the author of a book titled *A Scheme of Heaven: Astrology and the Birth of Science* concluded that "the methods of today's [big data] scientists are often uncomfortably close to those of astrology's ancient sages" (Boxer 2020, back cover).

Numerous predictor variables can be tested, and many versions of regression models can be run to produce findings that align with the sponsor's objectives. Thus, machine learning methods can facilitate the development of models to support prior hypotheses or can be used to produce "interesting hypotheses" after the analyses have been done.

In one of Scott's "Tom Swift" studies, Tom used standard procedures, starting with 31 observations and 30 potential variables. He used stepwise regression and only included variables with t-statistics greater than 2.0, which approximates to the commonly used statistical significance cut-off of $p \leq 0.05$. Along the way, he dropped three outliers. The final regression had eight variables and an R^2 (adjusted for degrees of freedom) of 0.85. Not bad, considering that the data were from Rand's book of random numbers (Armstrong, 1970).

With the large sample sizes that are available as a consequence of electronic transactions and record keeping – the phenomenon of "big data" – finding statistically significant correlations is trivial. As a consequence, researchers and the readers of their research findings can be deluded into thinking that the findings are meaningful. Statistics of fit and "significance" that are typically provided with models estimated with the statistical methods used in machine learning are confusing, even to senior researchers, and lead to overconfidence (Armstrong, 2012a).

A colorful recent illustration of the absurdity of applying machine learning to a large data set was a model of the price of Bitcoin. The model was estimated by providing the neural network procedure with time series data on the planets' zodiac signs from which to draw "predictor variables" and to estimate parameters. The resulting model provides a remarkable statistical fit to the volatile Bitcoin price data from 2010 to 2019 from which it was estimated (Boxer 2020, p. 252, fig. 10.5).

We expect that most readers of this book – we hope all – will not be rushing to estimate zodiac models to predict market prices in the hope of early retirement. Unfortunately, however, due to the illusion of relationships that statistical fit provides, there is no shortage of people willing to assume that experts' machine learning models have predictive validity. In practice, published "data models" – presumably intended to promote their developers' machine learning procedures – have failed to provide forecasts that are more accurate than those from simple models based on cumulative knowledge (Armstrong and Green, 2019).

The failure of machine-learning methods to pass the test of predictive validity is not surprising. They violate the scientific method, particularly in that they ignore cumulative scientific knowledge and theory on whatever it is they are attempting to model.

5.3.8   Using Faulty Logic

To assess scientists' logic, Mahoney and Kimper surveyed scientists in physics, biology, psychology, and sociology. One hundred each were randomly selected from professional directories in each field and 77 completed questionnaires were returned. *Over half the scientists did not recognize disconfirmation as being logically valid.* Moreover, while all of the physicists correctly rated denying the antecedent and affirming the consequence as invalid, only 64 percent of sociologists recognized those logical fallacies (Mahoney and Kimper, 1976).

In his examination of the K&T research on prospect theory, Gal (2006) concluded that the greater weighting of losses is an artifact of the well-known status quo bias and is not a fundamental psychological propensity. Smith pointed out that Adam Smith had identified, two centuries earlier, that the phenomenon "was derived from a more fundamental asymmetry between joy and sorrow" (2013, p. 293).

Gal and Rucker (2018a) concluded the K&T biases did not follow logically from K&T's research. To get further input on that question, they asked for commentaries by Simonson and Kivetz (2018) and Higgins and Liberman (2018). Those commentaries were followed by a reply from Gal and Rucker (2018b). For the most part, the commentaries were supportive of Gal and Rucker's conclusions.

5.3.9   Using ad hominem and Authority Arguments

Advocacy can also involve attacks on the credibility of researchers who test the status quo hypothesis against alternatives and claims that the status quo hypothesis is supported by the leading experts in the field. Aristotle noted that *ad hominem* arguments are not logical. For example, "Why should we take young Herr Einstein's ideas seriously? He is a mere patent clerk." Nevertheless, claims that those with different views are not qualified, are biased, or have dishonorable motives, and should therefore be excluded from academic and public debate are common.

We believe that, as a result of government involvement in science, such behavior has increased. Governments can, and do, bestow authority on particular hypotheses. For example, Miller (2007) lists six "unassailable paradigms" that researchers cannot challenge without being silenced and losing government grants. He included "Cholesterol and saturated fats cause coronary artery disease," "Human activity is causing global warming," and "even a tiny amount of a toxin, such as radiation or cigarette will harm some people."

As we describe in Chapter 6, the natural tendency to advocacy is increased by the incentives researchers face.

5.4 Prevalence and Acceptance of Advocacy

Mahoney's (1976) book on the practices of scientists states that "the average scientist tends to be … confident about his rationality … expertise … objectivity … and insight." But he found that objectivity in the search for truth was often violated in practice (see Mahoney, 1976, pp. 6–14).

As we noted in Chapter 2 – Defining the Scientific Method – there is a standard solution to the problem of advocacy. That solution is to use Multiple Reasonable Hypothesis Testing (MRHT).

To assess the objectivity of papers in a leading journal in management, one study coded all 120 empirical papers that were published in *Management Science* from 1955 to 1976. Of those papers, only 22 percent used MRHT. Sixty-four percent used advocacy, and 14 percent tested no hypotheses (Armstrong, 1979).

Another study used the same definition of advocacy – use of a single hypothesis – to code 1,700 empirical papers published from 1984 to 1999 in six leading marketing journals. The study found that only 13 percent used MRHT, while 74 percent of the papers used advocacy, and 13 percent had no hypotheses (Armstrong et al., 2001).

A paper in one of the most respected journals in management concluded that objectivity should be replaced by advocacy given that advocacy was widely practiced by the more than 40 "eminent physical scientists" who were interviewed in the author's study. The scientists described having sought support for selected hypotheses and actively avoiding and suppressing disconfirming evidence (Mitroff, 1972). The author subsequently stated, "We need better models of science that are based, if only in part, on what scientists actually do" (Mitroff and Mason 1974, p. 1501).

In a response to Mitroff's papers, Armstrong (1980c) "revealed" that "Mitroff" was the fictitious name for a group of scientists who wished to demonstrate that papers in violation of the scientific method could be published in a leading scientific journal. In effect, Armstrong used the advocacy approach and avoided mentioning disconfirming evidence – in particular, that he had met with Mitroff on a number of occasions. Yes, Scott's paper was a spoof, but the Mitroff papers were not.

5.4.1 Extent to which Psychology Papers Are Scientific

Once again, we turn to the Open Science Collaboration (OSC) data. This time we consider the extent to which 94 psychology studies complied with the scientific method.

We hired five undergraduate students to rate the 94 studies for compliance with science. They spent an average of 25 minutes per study on the task. Table 5.2 shows the percentage of studies that complied with each criterion, based on the modal rating of the five raters.

Although no study complied with all of the criteria, all studies complied with at least one criterion. On average, each paper complied with three criteria. Thus, while each paper had the potential to contribute to scientific knowledge, there was room for improvement in all of the studies.

Surprisingly – especially given that OSC papers were chosen because they described experiments – only 2 percent fully complied with

Table 5.2. Percentage of psychology studies compliant with science (modal rating of five novice raters)

1. Important problem	25.5
2. Comprehensive literature review	0.0
3. Sufficient disclosure	7.4
4. Objective design	46.8
5. Valid and reliable data	42.6
6. Valid simple methods	24.5
7. Experimental evidence	2.1
8. Logical conclusions	61.7

the requirement to test hypotheses with proper experiments (criterion 7). A glance at Checklist 1 suggests why that occurred – the papers did not provide comparisons on the accuracy of predictions from each hypothesis by using out-of-sample data (Criterion 6). That is, they did not test for predictive validity.

The reliability of the undergraduate students' ratings was modest, but better than could be expected from chance agreement alone. In particular, four or five of the five raters agreed on 46.1 percent of the ratings.

5.4.2   Advocacy vs Objectivity among Psychology Papers

We expected that the primary reason for the lack of compliance with science would be advocacy. Forty-four of the 94 OSC studies (46.8 percent) used an objective design, as judged by our novice raters in their rating of criterion 4. The other studies were coded as "advocacy."

Advocacy – the absence of objectivity – can be advanced by the failure to comply with other than just the MRHT criterion (Criterion 4). Based on modal ratings, the advocacy studies complied with a median of one criterion (mean of 1.4 criteria), compared to three criteria (mean of 2.9) for the studies rated as having an objective design. Table 5.3

Table 5.3. Advocacy studies' compliance with science (Modal rating of five novice raters – percentage of compliant papers)

Criterion	Objective (n = 44)	Advocative* (n = 50)
1. Important problem	27	24
2. Comprehensive literature review	0	0
3. Sufficient disclosure	11	4
5. Valid and reliable data	50	36
6. Valid simple methods	30	20
7. Experimental evidence	2.3	2.0
8. Logical conclusions	68	56

* Papers coded as objective or advocative based on majority of raters

omits criterion 4 (objective design) so as to compare the ratings on the other criteria for studies rated as objective with those studies that were not.

The studies rated as objective rated higher than the advocative studies on all the other criteria except the comprehensive literature review criterion. None of the OSC studies were rated as compliant.

The ratings of the studies against the criterion of drawing logical conclusions (Criterion 8) is noteworthy. Compliance generally was high relative to the other criteria, but whereas only a simple majority of advocative studies were rated as having drawn logical conclusions, a more than two-thirds majority of objective studies were rated as having done so.

6 SCIENTIFIC PRACTICE: PROBLEM OF JOURNAL REVIEWS

Publishing important and useful papers has been a fulfilling role for the founding editors of journals, and scientists were thankful to be able to publish research on topics that they regarded as important. The process seemed to work well for centuries.

Up to the mid-1900s, only a small percentage of university faculty members published scientific papers. Most faculty were concerned with teaching.

The success of government sponsored military research during World War II (WWII) encouraged the US government to fund research at universities after the war. The success of academics is now judged primarily on the dollar value of grants obtained and the number of papers published in journals that employ mandatory peer review. Peer reviews are dominated by the opinions of academics whose opinions are compatible with those of the journal editors. Not surprisingly, the number of papers has proliferated, as has the number of journals in order to accommodate the volume of papers.

Concern with pleasing reviewers in order to get papers published led to what Scott termed the "Author's Formula" for authors who aim to publish many papers:

> Authors should: (1) not pick an important problem, (2) not challenge existing beliefs, (3) not obtain surprising results, (4) not use simple methods, (5) not provide full disclosure, and (6) not write clearly. (Armstrong 1982a, p. 197.)

The peer review arrangements that are now standard became institutionalized during the twentieth century (Burnham, 1990). Journal editors no longer have time or specialized knowledge to review every paper themselves, so papers are assigned to associate editors. They, in turn, select academics to review the papers. As a result, reviewing is often outsourced to people with lesser expertise, and without the overriding objective of publishing useful scientific papers.

Typically, two or three unpaid reviewers, each spending an average of two to six hours per paper, completed the reviews (Jauch and Wall, 1989; King et al., 1981; Lock and Smith 1990; Yankauer 1990). While the reviewers sometimes made suggestions for improvements, their main role was to recommend whether a paper should be published. Scott Armstrong's (1997) review of research on peer review found that, in practice, innovative or controversial research tended to be blocked by the process.

A few years ago, Scott attended a small conference consisting of leading marketing science authors and editors. The aim of the conference was to find ways to publish papers that are useful for practitioners.

The conference began with a video of top marketing executives in large US corporations. When asked whether they read the leading academic journals, the executives all replied "no." Why? Because nothing in the papers seemed useful to them.

The editors of the leading marketing journals at that conference were asked what they thought of the executives' assessments, and they replied that they agreed. But why did they accept useless papers? The editors replied that they had to accept the papers because the reviewers could find no major errors.

In short, the publishing system shifted away from the editors' focus on publishing useful scientific findings, to reviewers' assessments of whether a paper contains mistakes. As a consequence, the reviews are poor at identifying useful scientific papers, and at rejecting those that are not.

We provide solutions to the problems created by journal mandated peer reviews for authors in Chapter 10 (How Scientists Can Disseminate Useful Findings) and for journal editors in Chapter 11 (How Stakeholders Can Help Science) in a section on "Scientific Journals." For example, in contrast to journal-mandated reviews, reviews solicited by the author from researchers working in areas related to the topic of their paper *do* help to improve the paper.

6.1 Failure to Find Errors

How carefully do reviewers examine the papers that they review?

As we have discussed, contributing to science depends on building on prior knowledge, yet an audit of papers in three medical journals concluded that, "a detailed analysis of quotation errors raises doubts in many cases that the original reference was read by the authors" (Evans et al., 1990).

Another study examined the number of substantive mistakes in 50 randomly selected citations in each of three issues of public health journals. Thirty percent of the citations incorrectly summarized the cited papers' findings; half of those erroneous summaries bore no relationship to the cited papers' findings (Eichorn and Yankauer, 1987).

A fictitious paper with 10 major and 13 minor errors was sent to all 262 reviewers for the *Annals of Emergency Medicine*; 203 reviews were received. On average, the reviewers identified only 23 percent of the errors. Even more worryingly, 68 percent of the reviewers failed to notice that the paper's conclusions had no logical connection to the evidence that it had presented (Baxt et al., 1998).

In a more recent study, 607 medical journal reviewers read a fictitious paper containing nine important errors. On average, each reviewer found only 29 percent of the errors (Schroter et al., 2008).

As we described earlier, Armstrong and Overton (1977) developed a simple and effective way to estimate non-response bias in mail surveys. The paper had been cited over 15,000 times by late 2019. Yet, in a sample of 50 papers published in top journals that cited the procedure, only one accurately represented the procedure. Most of the citations used it to support an incorrect procedure for dealing with non-response bias, suggesting that the authors of those papers had not read the Armstrong and Overton paper. In addition, many of the citing papers made mistakes in the references, such as incorrect spelling, many of which were identical to the mistakes made in earlier citations of the paper (Wright and Armstrong, 2008).

In short, mandatory peer review is not effective for finding errors.

6.2 Advocacy by Reviewers and Editors

Scott had a phone call from a researcher whom he did not know. The caller was distressed because he had been told that a journal would accept his paper on the condition that he remove a citation to

Scott's work that was central to his argument. Scott responded that he would understand if the researcher chose to comply with the journal editor's request and thanked him for his call.

A 1980 survey of 311 psychology and statistics academics about their most recent revised and published article found that 76 percent claimed to have been pressured to conform to the subjective preferences of a reviewer. As many as 73 percent had encountered false criticisms, and 8 percent made changes that they knew were wrong in order to satisfy a reviewer (Bradley, 1981).

More recently, and encouragingly, as we described earlier, the editor of the *Annals of Internal Medicine* bravely resisted efforts by experts in the field to suppress a set of reviews of the evidence on the health effects of red and processed meats (Laframboise 2020a, 2020b). The comprehensive reviews concluded that eating meat was *not* harmful to health (Carroll and Doherty, 2019).

Reviewers blocking the publication of papers with findings that challenge their own research or beliefs appears to be common. A survey of 60 leading economists, including 15 Nobel Prize winners, surveyed by Gans and Shepard (1994), concluded that mandatory peer review tends to result in the rejection of papers with important scientific findings that challenge the status quo.

To test confirmation bias among reviewers, Mahoney (1977) sent two versions of the same fictitious paper to 75 reviewers for the *Journal of Applied Behavior Analysis*. Version 1 described findings that supported a commonly accepted hypothesis while version 2 – which used the same methods – reported contrary findings. Twenty-four reviewers responded. The confirmatory version 1 received an average rating of 4.2 out of 6 – higher ratings meaning more likely to be accepted for publication on the basis of the methodology – from 10 reviewers, while the 14 reviewers who rated divergent version 2 scored it at 2.4 on average. The reviewers' "summary recommendations" on whether to accept or reject the papers were similarly split; version 1 received an average rating of 3.2, whereas version 2 was rated at only 1.8.

Similar experimental findings were obtained in psychology by Goodstein and Brazis (1970), Abramowitz et al. (1975), and Koehler (1993), and in biomedical research by Young et al. (2008).

Another problem is coercive citation (Martin, 2013). Authors can be pressured to cite a reviewer's papers and to cite papers that were previously published in the journal.

6.3 Unreliable Reviews

A 1982 special issue of *Behavioral and Brain Sciences* examined peer review by using a target article by Peters and Ceci (1982). Peters and Ceci had resubmitted 12 papers – changing titles, authors, and affiliations – to the same 12 prestigious psychology journals that they had been published in 18–32 months earlier. Of the 38 editors and reviewers, only three (8 percent) detected the resubmissions. As a consequence, nine of the 12 articles were reviewed again. Eight of them were rejected.

More than 50 commentators examined aspects of the study. In his commentary, Scott asked a convenience sample of 21 full professors to predict how many of the resubmissions were (1) detected by the journals as a prior publication and (2) of those not detected, how many were rejected. The professors *predicted* that 66 percent would have been detected and, of the undetected, that 42 percent would have been rejected. In reality, only 25 percent of the 12 journals detected that the paper had been previously published and 89 percent of the undetected papers were rejected (Armstrong, 1982b).

Earlier versions of Peters and Ceci's paper had been rejected by *Science* and by the *American Psychologist*. In apparent contradiction to those earlier assessments of the value of this paper, it proved to be of substantial interest to researchers, amassing more than 1,200 citations by mid-2020. Despite the paper's success, Peters was treated poorly by his department for having published it (Ceci, 2020).

Reviewers have a low level of agreement among one another because they depend heavily on their opinions, rather than on experimental evidence. Many studies have found low reliability among reviews. See, for example, a summary of research on peer review for manuscript and grant submissions by Cicchetti (1991).

Frey (2003) found that among reviews in psychology, journals rejected two out of every three papers if reviewers were split in their publication recommendations. In effect, a single reviewer often has the power to reject a paper.

6.4 Failure to Improve Papers

Authors often pay little attention to reviewers' suggestions when their papers are rejected. Wilson (1978) found that 85 percent of the papers rejected by the *Journal of Clinical Investigation* were eventually published elsewhere, and the majority of these were either not changed or changed in only minor ways.

Similar findings were obtained by an examination of 61 papers rejected by another journal before they were submitted to the *American Journal of Public Health*. Forty-eight percent had not been revised (Yankauer, 1985). A study of papers rejected by the *American Political Science Review*, concluded that of the 263 papers that were then submitted to another journal, 43 percent contained no revisions based upon the reviews the authors had received (Patterson and Smithey, 1990).

Lock and Smith (1986) and Garvey et al. (1970) found similar patterns for papers rejected by the *British Medical Journal* and for papers in the social sciences, respectively.

6.5   Fooled by Fraudulent Reviews and Bafflegab

As long as there are incentives to publish papers, researchers will find ways to do so. For example, when given an opportunity to suggest reviewers, authors have provided fictitious contact information that directs reviews back to themselves, thereby allowing them to submit glowing reviews for their own papers (Gao and Zhou, 2017).

Mahoney (1976, p. 85) gave sarcastic advice to researchers whose only aim was to get published: "Whenever you have a choice between common language and technical argot, use the latter." Authors who use clear writing do so at personal risk, he wrote (Mahoney 1976, pp. 85, 96). Is there any evidence to support his advice?

As it happens, there is. "Doctor Fox" was an actor who looked distinguished and sounded authoritative. He had an impressive, but fictitious, biography and delivered a one-hour lecture on a subject about which he knew nothing. The talk, "*Mathematical Game Theory as Applied to Physician Education*," was delivered on three occasions to a total of 55 highly educated social workers, psychologists, psychiatrists, educators, and administrators. The lecture consisted of meaningless words, false logic, contradictory statements, irrelevant humor, and meaningless references to unrelated topics. It was followed by 30 minutes of discussion. According to a questionnaire completed after the talk and discussion, the audience considered Dr. Fox's lecture to be clear and stimulating. None of the subjects realized that the lecture was pure nonsense. The paper describing the study (Naftulin et al., 1973) was widely read and has over 600 citations to date.

Complexity impresses readers as well as listeners. A follow-up to the Dr. Fox study used abstracts from published papers in management, economics, and sociology journals. Alternative versions of each abstract were created, one more convoluted than the original, and one simpler than the original. A convenience sample of academics rated the abstracts; they judged the authors of the more complex versions as more competent than the authors of the simpler versions (Armstrong, 1980c).

Crichton (1975) analyzed the prose of a convenience sample of articles from three issues of the *New England Journal of Medicine* and identified 10 recurring faults. They were:

1. Poor flow of ideas
2. Verbiage
3. Redundancy
4. Repetition
5. Wrong word
6. Poor syntax
7. Excessive abstraction
8. Unnecessary complexity
9. Excessive compression
10. Unnecessary qualifications

He concluded that obfuscation was a "firm tradition" and – as Mahoney (1976) did the following year – speculated that plain writing would harm a scientist's career.

The abstract is the most important part of a paper. One study examined nearly 710,000 abstracts from 123 journals published between 1881 and 2015. Using two different measures of reading difficulty, the study found a strong trend in the direction of worsening readability over time. Having more authors was also associated with worse readability. The study concluded that "more than a quarter of the scientific abstracts now have a readability considered beyond college graduate level English" (Plavén-Sigray et al., 2017).

An analysis of 4,160 papers in the three leading finance journals published from 2000 to 2016 found that their writing was more complex than the writing of 8,236 papers published in other major finance journals; the median Flesch-Kincaid Index score was nearly a grade and a half higher. Moreover, among papers in the leading journals, citation rates *increased* with the difficulty of reading the paper (Berninger et al., 2018).

Another experiment found that readers rated explanations of studies' findings with irrelevant words related to neuroscience as more convincing than those without the irrelevant words (Weisberg et al., 2008). The effect was replicated in three further experiments by Weisberg et al. (2015). In a similar experiment, reviewers gave higher ratings to papers with irrelevant complex mathematics than to those without the irrelevant material (Eriksson, 2012).

In short, bafflegab helps authors with nothing to say to get published and cited as long as they advocate the beliefs that are consistent with those of reviewers and other researchers in the field. It is of course a deceptive practice that obstructs the accumulation of useful scientific knowledge.

6.6   Distracted by Statistical Significance Tests

For about a century, most leading social science journals insisted that quantitative papers must include tests of statistical significance. By 2007, statistical significance testing was included in 98.6 percent of published empirical studies in accounting, and over 90 percent of papers in political science, economics, finance, management, and marketing (Hubbard, 2016, ch. 2).

There is much evidence that such tests are harmful to scientific progress. (For a comprehensive review of the harm caused by the tests, see Ziliak and McCloskey's excellent 2008 book, *The Cult of Statistical Significance: How the Standard Error Costs Us Jobs, Justice, and Lives*.) When journal editors insist on statistically significant test results, researchers can be tempted to adopt unscientific practices in order to improve their chances of publishing a paper.

Most readers of scientific papers do not understand what statistical significance means. For one thing, they assume that it is a measure of the importance of the findings or the probability that the finding is true. It is neither, as can be understood by the fact that simply by increasing the sample size, statistical significance can be achieved for any study. Consider the study of more than 18,000 professional basketball games that concluded that teams that are behind by a point at half time are more likely to win than to lose the game (Berger and Pope, 2011). The finding was statistically significant, but of no practical significance for basketball teams. Would a coach really instruct players to miss a shot that could even the scores at half time?

Researchers also misunderstand statistical significance testing. A study of empirical papers in leading economics journals found that significance tests led authors to faulty conclusions (McCloskey and Ziliak, 1996). Furthermore, leading econometricians performed poorly when asked to interpret standard statistical summaries of regression analyses (Soyer and Hogarth, 2012).

Meehl, one of the most important psychologists of the twentieth century, concluded that "reliance on merely refuting the null hypothesis ... is basically unsound, poor scientific strategy, and one of the worst things that ever happened in the history of psychology" (1978, p. 817). Gigerenzer et al. (2004) provided a useful summary of the problems with the approach in their book chapter titled "The null ritual: What you always wanted to know about significance testing but were afraid to ask." Other scientists have reached similar conclusions.

Schmidt is one. He offered a challenge: "Can you articulate even one legitimate contribution that significance testing has made (or makes) to the research enterprise (i.e., any way in which it contributes to the development of cumulative scientific knowledge)?" (1996, p. 116). We are not aware of anyone who has been able to answer the challenge. Schmidt confirmed that was still the case in personal correspondence at the time of writing this book.

Because statistical significance is unrelated to importance, its use harms decision-making. In one study, 261 researchers who had published in the *American Journal of Epidemiology* were presented with the findings of a comparative drug test and asked which of the two drugs they would recommend to a patient. More than 90 percent of subjects presented with statistically significant drug test findings ($p < 0.05$) recommended that drug; in contrast, fewer than 50 percent of those who were presented with the same estimate of benefits, but statistically insignificant results did so (McShane and Gal, 2015). The same researchers show academics from a wide variety of fields including economics, psychology, and even statistics (McShane and Gal, 2017) make similar dichotomization errors around the 0.05 threshold, thereby ignoring effect size estimates.

A vivid example of how the use of statistical significance tests can lead to bad decisions is the "Right-turn-on-red" rule for automobiles. Several smallish studies on the rule concluded that the higher rate of crashes compared to the alternative rule was not statistically significant. As the rule became widely adopted and larger studies were conducted, however, it became clear that allowing automobiles to turn right on red was a major health hazard for pedestrians and bicyclists (Hauer, 2004). Rather than defer to statistical significance tests, policy makers would have been better to conduct a cost–benefit analysis on the basis of the studies' estimates of increased accidents.

In practice, testing for statistical significance harms progress in science. For example, one study used significance tests to conclude that combining forecasts did not improve predictive validity (Koning et al., 2005). In fact, combining predictions from different models and methods is arguably the most effective method for increasing predictive validity (Armstrong, 2007; Armstrong and Green, 2018).

Evidence on the harm due to statistical significance testing is such that the US Supreme Court, in a case involving the effectiveness of a drug (*Matrixx* v. *Siracusano)* reached a unanimous decision that "statistical significance" cannot be used as evidence of efficacy under any conditions (Ziliak 2011). The court decision is likely to have a major beneficial impact. For example, advertisers who use statistical significance might be sued for misleading advertising by consumers or competitors. Might journals, too, be held liable for damages if they insist that authors use tests of statistical significance?

6.7   Blind to Importance of Authors' Previous Contributions

Double blind peer review is the standard practice to remove potential bias in journal publication decisions. Reviewers do not know the name, sex, education, or experience of the authors, and vice versa. Does that help?

The problem is that removing the name of the author erases important information about whether the author had previously made useful scientific discoveries. It is widely accepted that the most successful scientists – a tiny percentage of all researchers – contribute the majority of *useful scientific findings*.

Track records matter. Consider that you have to choose among surgeons to perform a risky operation on your child. Would you want to know which of them had the most success in performing the operation? Similarly, how could it fail to be useful to know if a scientist has previously published papers with useful scientific findings?

6.8   Long Delays in Publication

Reviewers seldom complete reviews in a timely manner, authors are slow to make revisions, and many papers go through more than one revision. Björk and Solomon (2013) estimated the delays for 2,700 papers in 135 journals covering nine areas of study: on average, it took 6.4 months for a paper to be accepted and another 5.8 months to be published. The total delays varied among fields, ranging from about nine months for chemistry, engineering, and biomedicine, to 18 months for business and economics.

Also, given that journals only allow authors to submit to one journal at a time and that acceptance rates are low for most journals, a typical paper can spend years in review at a succession of journals before being published.

7 SCIENTIFIC PRACTICE: PROBLEM OF GOVERNMENT INVOLVEMENT

> [T]he NSF has also done harm to the progress of ... science ... First, by converting "grantmanship" into as important an ingredient of professional success as original and penetrating contributions ... Second, by channeling ... research into directions favored by dispensers of NSF funds.
>
> Milton Friedman (1981 "open letter" to the president of the National Academy of Sciences quoted in Friedman, 1994)

In this chapter, we raise making a useful contribution as a motive for university researchers, discuss some of the recent history of government involvement in science, and examine the evidence on whether or not government funding and regulation of research is beneficial.

We provide solutions to the problems created by government involvement in science in Chapter 9 (How Scientists Can Discover Useful Knowledge) in subsections titled "If you need funding, ensure that you will nevertheless have control over your study," and "Consider role-playing as an alternative to experimentation." In Chapter 11 (How Stakeholders Can Help Science) in a section on "Governments, Regulators, and Courts" we make suggestions for how governments can best help the quest for useful scientific knowledge.

7.1   Being Useful

University researchers were originally expected to do research that addressed important problems. They had much freedom to decide what research to do. Private consulting was allowed but was frowned upon and limited. At the Wharton School in the 1960s, for example, professors had to receive permission from school managers to consult to businesses or governments, and could not do so for more than a day each week during the school year.

In January 1946, the federal government created the Research Grants Office. More than 75 years later, scientists now spend much time applying for grants, and universities reward professors if they obtain grant money from the government or other funders. In short, university managers encourage consulting to the government, and they receive substantial contributions to university overheads in return.

The US Federal Government is insistent that colleges and universities apply for grants. If they decline research money, they must also decline government-sponsored student loans and other programs. At the time of writing, there were only 18 US colleges or universities that refused government funding (Clancy, 2017; updated 2020). In Australia, Canada, and the UK, public universities predominate with few private universities of any size.

Politicians, selected experts, and interest groups propose topics that they consider to be promising. To be "fair" to researchers, open appeals are made to bid for grants. Peer review is used to decide which proposals to fund. In more recent times, there has been an emphasis on demographically diverse groups of researchers. However, a meta-analysis of 93 studies on team performance found no benefit from demographic diversity (Stewart, 2006).

Scientists have raised concerns that government involvement in research would be harmful ever since governments first started increasing their roles. For example, Leo Szilard, the scientist who in late 1939 wrote the letter that led to the Manhattan Project (to develop the atomic bomb), used a short fictional story about the Mark Gable Foundation to indicate how he expected things to turn out when governments get involved in funding science:

> "I have earned a very large sum of money," said Mr. Gable . . .
> "And . . . I want to do something that will really contribute to the happiness of mankind . . ."
> "Would you intend to do anything for the advancement of science?' I asked.
> "No," Mark Gable said. "I believe scientific progress is too fast as it is."

"[T]hen why not do something about the retardation of scientific progress?"

"That I would very much like to do..."

"Well," I said, "... You could set up a foundation, with an annual endowment ... Research workers in need of funds could apply for grants ... Have ten committees ... appointed to pass on these applications. Take the most active scientists and make them members of these committees. And the very best ... should be appointed as chairmen ... Also have ... prizes ... for the best scientific papers of the year."

"First of all, the best scientists would be removed from their laboratories and kept busy on committees ... Secondly, the scientific workers in need of funds would concentrate on problems which were ... pretty certain to lead to publishable results ... [P]retty soon science would dry out. There would be fashions. Those who followed the fashion would get grants. Those who wouldn't ... would learn to follow the fashion, too." Leo Szilard (from "The Mark Gable Foundation," 1961).

7.2   Funding

To compel a man to furnish contributions
of money for the propagation of opinions
which he disbelieves and abhors, is sinful and tyrannical.
        Thomas Jefferson (1779)

Because we agree with Jefferson's principle, we have never applied for government research grants. We also question whether chasing government grants is an efficient approach if the objective is to conduct useful scientific research.

Adam Smith wondered why Scotland's relatively few academics were responsible for more scientific advances during the Industrial Revolution than England's more numerous academics. At the time, professors in Scotland were paid by attracting fee-paying students and by businesspeople and others who were persuaded that the professors were doing useful research on important problems. On the other hand, the enormous endowments from kings and others available to Oxford and Cambridge professors removed from them the need to conduct useful research (Kealey, 1996, pp. 60–89). Modern universities around the world resemble those of eighteenth-century England, with external support and direction undermining scientists' motivation to conduct useful scientific research.

7.2.1 Government Funding since WWII

Government funding of research was high during WWII. Toward the end of 1944, US President Roosevelt commissioned a report from the Director of the Office of Scientific Research and Development, Vannevar Bush, to provide recommendations on, among other things, what the government could do "to aid research activities by public and private organizations" (President Roosevelt's letter in Bush, 1945). Note that the request implies advocacy for government aid was expected.

In his report, Bush wrote:

> New impetus must be given to research in our country. Such impetus can come promptly only from the Government. Expenditures for research in the colleges, universities, and research institutes will otherwise not be able to meet the additional demands of increased public need for research.
>
> Further, we cannot expect industry adequately to fill the gap. Industry will fully rise to the challenge of applying new knowledge to new products. The commercial incentive can be relied upon for that. But basic research is essentially noncommercial in nature. It will not receive the attention it requires if left to industry. Vannevar Bush (from "The special need for federal support," 1945)

After the war ended, the US government – presumably motivated by the opinions Vannevar Bush expressed in his 1945 report in response to President Roosevelt's request – claimed that the private sector would not do enough "basic" research to maintain the US's economic and military leadership – and extended funding to research that had no clear defense objective. Government research funding was made available to universities as well as to the private sector, with the aim of increasing the amount of research that was done.

The new involvement of government in funding research left funding agencies with the question of how to choose which research proposals to fund. Without the discipline that goes with investing one's own savings, there is little reason to assume that research-funding decisions contribute to economic development. Government funding agencies further reduced the likelihood that funds would be allocated to the most useful projects by using the fairness principle to choose grant recipients on the basis of, for example, demographic characteristics, ahead of any consideration of importance of the topic and the likely performance of the researchers.

The demand for researchers in universities grew rapidly as government funding increased. With no reason for the supply of self-motivated scientists to increase, extrinsic rewards were used to attract additional researchers. For example, between 1965 and 1972 the US National Science Foundation awarded US$230 million of "Science Development" funding to 31 "second tier" universities in the hope that they would become "centers of excellence." They hired more researchers and published more papers, but Brush (1977) concluded from a review of the program that "these findings certainly would not be likely to persuade Congress to launch another Science Development programme" (p. 396).

As we will discuss in Chapter 8 (What It Takes To Be a Good Scientist), money is not the primary motivation for researchers who do useful scientific research. Moreover, governments' contributions to universities' overheads came with the expectation that universities hire researchers to meet demographic diversity targets without any evidence that more demographic diversity among scientists would produce more useful scientific discoveries.

In 1940 – five years before the Bush report, and before the United States became involved in WWII – federal spending on research and development (R&D) amounted to US$97 million, or 0.1 percent of GDP. Eight years after the report, in 1953, the figure was US$2,200 million (National Science Foundation, 1953). Even after adjusting for inflation, federal R&D expenditure increased more than twelvefold over the 13 years. As a share of GDP, federal R&D spending increased sixfold over the period.

Dwight D. Eisenhower was sworn in as US President in 1953, when federal R&D spending had reached 0.5 percent of GDP. Eight years later – in the 1961 fiscal year in which he left office – spending had increased to 1.6 percent of GDP (Office of Management and Budget, 2018, p. 190, tab. 9.7). The scale of government involvement in science led President Eisenhower to warn about its consequences in his 1961 farewell address to the nation, as follows:

> The prospect of domination of the nation's scholars by Federal employment, project allocations, and the power of money is ever present – and is gravely to be regarded. Yet, in holding scientific research and discovery in respect, as we should, we must also be alert to the . . . danger that public policy could itself become the captive of a scientific-technological elite. Dwight D. Eisenhower (from *Farewell address to the nation*, 1961)

Following Eisenhower's warning, federal R&D expenditures continued to increase in real – inflation adjusted – terms for a further six years to 1967 – 1965 being a slight exception. As a share of GDP, however, spending peaked in 1964 at 2.1 percent of GDP, and declined thereafter to return in 2017 to a level not seen since the middle period of Eisenhower's term as United States President of 0.6 percent of GDP. At the time of writing, spending is estimated to remain at that level through to the fiscal year 2020 (Office of Management and Budget, 2018, p. 190, tab 9.7).

In a sense, then, Eisenhower's warning was heeded, albeit belatedly, and then slowly over a period of more than 50 years. Nevertheless, spending remains high at six times the percentage of GDP that it was in 1940, before the United States entered WWII.

Do politicians, government officials, and expert panels have the knowledge or incentive to identify the most useful research projects? Given that they are spending other people's money, can we be confident that the research will be cost-effective?

Evidence from natural experiments was summarized by Kealey (1996). He concluded that government-funded research was less useful than privately funded research and tended to crowd out private research.

Bush's (1945) assertion that firms would not do enough "basic" research to support long term gains in productivity was not supported by an analysis of manufacturing firms from 1948 to 1966, which found firms benefited from productivity gains by undertaking basic research. So why wouldn't they do it? Moreover, a survey of 119 firms accounting for around one-half of US industrial R&D expenditure found that their basic research expenditure declined from 5.6 percent to 4.1 percent of total research expenditure between 1967 and 1977 following the peaking of government R&D expenditures. Over the same period, projects aimed at "entirely new products and processes" declined from 36 to 34 percent and those with "less than 50-50 estimated chance of success" from 28 to 25 percent. The respondents explained that such research projects were less profitable due to increased government regulation (Mansfield, 1980).

So, with the greatly increased government funding of university researchers, what percentage of new products and processes in the WWII are the result of academic research? Survey responses from 77 major firms for the period 1986 to 1994 concluded that 11 percent could not have been developed without knowledge of recent academic research, at least not without a long delay. Respondents indicated that 7 percent of new products and processes had been developed with very substantial help from recent academic research (Mansfield, 1998). These are not impressive figures.

We, too, are unconvinced that the enormous diversion of resources to government approved research projects over nearly 70 years has delivered greater prosperity and security than would otherwise have been the case. That is also the conclusion that is suggested by an econometric analysis of the effects of R&D conducted by government-funded research institutes and by business on economic growth in OECD countries, and a review of the literature on the effect of R&D on productivity growth. The estimated effects (elasticities) in the econometric study were substantially negative for government and substantially positive for private R&D (OECD, 2003, pp. 84–85). A later review of evidence concluded that:

> On the basis of the evidence considered, privately financed R&D in industry should be treated as an investment and included in the relevant R&D stock. Returns to R&D are very high, but these high returns accrue only to privately financed R&D. Many elements of university and government research have very low returns, overwhelmingly contribute to economic growth only indirectly, if at all, and do not belong in investment. Sveikauskas (2007, p. 44)

In contrast to Vannevar Bush's conclusion that government spending on research is needed for peace-time economic development in advanced nations, our conclusion is that left to their own devices, scientists in private corporations and think-tanks would continue to do useful scientific research.

7.2.2 Government Funding and Advocacy

Government funding for research encourages advocacy that favors the preferred policies of those in power and of the scientists who control the grant committees. For example, in the United States, the Environmental Protection Agency (EPA) funds research on the dangers of human emissions of various kinds. Two studies on the effects of microscopic particles in the air were commissioned by the EPA (Dockery et al., 1993; Styer et al., 1995). The Dockery paper claimed that PM2.5 particles were killing people. The Styer paper, on the other hand, looked at PM10 particles – a particle size classification that includes the smaller PM2.5 particles – and could find no association with increased deaths. The Dockery paper has been cited nearly 10,000 times, whereas the Styer paper had been cited only 149 times by December 2021. The EPA sided with the Dockery study and funded further research on the effect of air quality on health by Dockery and colleagues. The Styer researchers got no further funding from the EPA.

In 1994 a US Government Accounting Office evaluation recommended improvements to the processes used by the National Science Foundation for reviewing research grant proposals. The report's authors were concerned that proposals were selected for reasons other than the extent to which they met stated criteria, and that reviewers were not rating to the same standard. An experiment to test whether those concerns were justified found that the overall ratings of 70 proposals by one of four reviewers were inconsistent with that reviewer's ratings against the four official criteria. The other three reviewers were consistent between their overall ratings and their criteria ratings, but each weighted the criteria quite differently. Another experiment found that the mean ratings of nine review panelists using a five-point-scale – which would be roughly equal if they were rating to the same standard – had a range of nearly one (Arkes, 2003).

Because government research funding typically comes with the expectation that the findings will support the funding agency's agenda, the research is likely to fail the criterion of objectivity. Without objectivity, it is not safe to draw conclusions about what policies would be beneficial. As we noted earlier, scientists are self-motivated to make useful scientific discoveries, and know better than remote policy makers and their agents what problems to address and how to design research in order to improve knowledge on how to deal with those problems.

Often a US government agency has a policy that they want to support with research. For example, in recent decades, much research has been conducted in order to address "equity" among different demographic groups. The purpose of that research is to find ways to transfer resources from one group to another in order to reduce inequality of outcomes. To achieve that ideal, it is believed by those who advocate it that past grievances must be addressed. Francis (1986, p. 102) discussed this movement, that started, it seems, in Australia in the 1980s.

Grievance studies, typically published in social science journals, focus on differences between demographic groups. They cast the differences as unfairness or inequality regardless of whether the differences arise from free choice. The government encourages researchers to "celebrate diversity." This is done by discussing the ways in which people differ. Unsurprisingly, some groups of people tend to have higher money incomes or wealth than others, and those differences are portrayed as unfair.

Scholars have questioned the value and consequences of the grievance studies. It seems unlikely that many such papers would be published without government funding for research on the topic.

To assess the worth of grievance studies papers, three researchers at Portland State University tested whether fake advocacy papers, consistent with grievance studies' themes, would be published in academic journals (Lindsay et al., 2018). Their papers started with a conclusion that was consistent with the goals of each journal to show that some demographic groups were being treated unfairly. The authors then used complex irrelevant arguments that were absurd and morally repellent. The arguments were drawn from published sources, including Hitler's *Mein Kampf*.

The hoax was cut short after ten months due to publicity about one of the studies. Of the 20 papers already submitted, four had been published online and three others had been accepted. Two were "revised and resubmitted," and one was a "reject and resubmit." The peer reviews were very favorable. *Gender, Place and Culture* honored a paper as one of the 12 leading papers in feminist geography over the first quarter century of the journal. In all, their record would have been in a range that qualified the authors for tenure.

The problem of intergroup conflict, which is at the crux of grievance studies, has been the subject of *scientific* studies for decades. And so, apparently unbeknown to those conducting grievance studies, the solution is already known; creating common objectives overcomes the negative effects of intergroup conflict. The famous Robber's Cave field experiment (Sherif et al., 1961) exemplifies those studies. The uniting power of common objectives was recently illustrated by the Oscar award-winning 2018 film, *Green Book*, which was based on real events. In the movie, an Italian American man who was prejudiced against black people was out of a job, and so agreed to chauffer a black pianist on a tour of southern US states at a time when there was much discrimination based on skin color. The chauffer's job required that he protect his client. The common goal of a safe tour was enough to unite the pair, who became lifelong friends.

Countries that emphasize the commonalities of its citizens would be expected to reduce conflict among citizens. For example, in Switzerland citizens can choose to live in a canton that most suits their preferences and situation. Frey's (2010) book, *Happiness: A Revolution in Economics*, describes how that freedom leads to increased life satisfaction. Similarly, the states and counties of the United States have different laws and regulations, and that allows people to choose to live in a community that they consider is best for them (see Murray, 2016).

7.2.3   Government Funding of Groups of Scientists

Government grants are generally bestowed on groups of researchers. Universities are keen to have teams of researchers that include researchers with different demographics in the proposals. The rewards are substantial for researchers who can obtain a stream of grant revenue to support many researchers: the higher the cost of the research, the larger the contributions to university overheads.

As a consequence, undertaking grant-funded research in teams has become the norm. In the physical sciences – broadly defined to include medicine, engineering, mathematics, and computer science – papers with multiple authors accounted for 53 percent of papers in the period from 1955 to 1959, compared with 85 percent in the period from 1996 to 2000. Corresponding figures for the social sciences were 16 percent and 45 percent. Among co-authored papers, the mean average number of authors increased from 2.6 to 4.2 in the physical sciences and from 2.3 to 2.8 in the social sciences (figures calculated from data provided in the "Supplementary Table S1," in Wuchty et al., 2007).

Do teams with more researchers *produce more useful scientific knowledge* than those who publish individually? Wu et al. (2019) addressed that issue in their analysis of more than 65 million papers, patents, and software from 1954 to 2014. They concluded that the important ("disruptive") scientific discoveries are produced disproportionately by individuals and by very small groups of researchers. They also found that when individuals and small groups were funded by government grants, they were no more productive than large groups in terms of important discoveries. If individual researchers who might otherwise have made useful scientific discoveries are being diverted into grant-getting activities – especially when they are expected to assemble teams that meet a grant committee's criteria – the discovery of useful scientific findings is likely to be harmed.

It is difficult to think of major scientific advances from groups. Research on groups found that creativity is low compared to individual creativity (Armstrong, 2006). People in groups are inclined to cooperate and agree, whereas science requires skepticism.

Researchers who depend on funding for income and career progression are unlikely to present findings that conflict with their funders' preferences and interests. As a consequence – whether willingly or reluctantly – they become advocates for the funders' beliefs and causes.

Arctic expeditions of discovery provide a vivid illustration of the difference that the source of funding can make to costs and outcomes. Karpoff's (2001) comparison of the 35 public Arctic explorations with the 57 private ones for the period from 1818 through 1909 found that the public ones were, on average per expedition:

1. More expensive (70 crew members needed for public vs 16 for private, and loss of 0.53 ships – 198 tons – for public, compared to 0.24 ships – 60 tons – for private);
2. Less healthy (e.g., 47 percent of members contracted scurvy for public, compared to 13 percent private);
3. More dangerous (death rate of 5.9 people per expedition for public, compared to 0.9 for private).

Karpoff concluded that the poor performance of the government-funded expeditions was due to perverse incentives, slow adaptation of important innovations, ineffective organizational structures, and poorly motivated leaders.

7.3 Regulation

Historically, scientists have tried to design studies that avoid harm because they realize that ignoring the natural concern for the welfare of others would not only be wrong, it would lead to disgrace and exposure to lawsuits brought by harmed research participants. They are the most likely people to be aware of any risks involved in their research and to understand how to best design studies to minimize those risks.

Regulations of the practices of scientists can impose enormous costs, yet they are typically imposed on the basis of the opinions of politicians and regulators who are not so well placed as individual researchers to design safe and effective research projects. For example, Hauer (2019) found that road safety regulations based on expert opinions led to unnecessary deaths.

While scientific evidence is often claimed to be the basis of regulators' opinions, government regulators often refuse to provide full disclosure (Cecil and Griffin, 1985). In a recent step to rectify those failures to comply with the scientific method, US Congressman Lamar Smith proposed changes in H.R.4012 – Secret Science Reform Act of 2014 (see Committee on Science, Space and Technology, 2019).

7.3.1 Consent Forms

Government officials began to regulate science with the aim of protecting subjects in experiments. Many scientists were upset by the presumption that they cannot be trusted to design experiments that protect their subjects. In contrast to scientists, who have much to lose if they harm subjects, when regulations harm people the typical solution is to hire more regulators and add more regulations.

Researchers are required to obtain signed "informed consent forms" from research participants in order to comply with research regulations in the United States and elsewhere. National regulations on informed consent were developed in order to meet the requirements of international agreements that originated with the Nuremberg Code of 1947, which was itself developed in response to unethical experimentation on prisoners in government established and run concentration camps (see e.g., Nijhawan et al., 2013 for a description of the historical origins of these requirements).

Was there a need for informed consent forms?

Governments relied on *examples* of studies that harmed subjects in order to justify regulations. Among these were the Tuskegee syphilis experiments; a radiation study where prisoners, mentally handicapped teenagers, and newborn babies were injected with plutonium; and the eugenics experiments in the early 1900s. Tellingly, these were all government experiments. It would be repugnant and personally and professionally dangerous for individual scientists to conduct such unethical projects without government support.

Regulators rely on speculations about what might possibly go wrong to insist on cumbersome informed consent protocols. Pronovost's lifesaving research on checklists in hospitals – discussed in Chapter 1 of this book – provides an example. Pronovost had tested the checklist at Johns Hopkins and, to find out if the checklist would also help reduce deaths in other and different hospitals, obtained the cooperation of the Michigan Hospital Association to test it in their hospitals. The use of the checklist saved perhaps 1,900 lives in the process of testing in those hospitals. And then someone complained that the experiment was being conducted without the use of informed consent forms. As a result, Pronovost and the Hospital Association were ordered to stop collecting research data ... but were not ordered to stop using the checklist.

7.3.2 Institutional Review Boards and Ethics Committees

The US Congress passed the National Research Act in 1974. It required that Institutional Review Boards (IRBs) would license and monitor research with human subjects. Nearly all researchers in institutions that receive federal funding must have their studies' topic, design, and reporting approved by an IRB if the study involves human subjects. That applies *even when the researcher receives no government funding* (Schneider, 2015, p. xix).

Equivalent bodies in the United Kingdom are called Research Ethics Committees, and in Australia are called Human Research Ethics Committees (HRECs). HRECs in universities and other institutions in which research involving human participants is conducted are required to ensure compliance with the guidelines in the 116-page *National Statement on Ethical Conduct in Human Research 2007* (National Health and Medical Research Council, the Australian Research

Council and Universities Australia, and Commonwealth of Australia, updated 2018).

What's not to like?

For one thing, the procedure is often time-consuming and expensive. In addition, the required changes can be onerous. When Harvard University graduate student, Kimberly Sue, wanted to interview female prisoners about opiate addiction, the Harvard IRB placed bizarre restrictions on her study, as described in "You Can't Ask That" (Schrag, 2014).

Neither Schneider (2015) nor Schrag (2010) could find evidence of serious harm by individual scientists in their reviews. For example, Schrag's analysis of a study of 2,039 research projects conducted between July 1974 and June 1975, found only three projects that reported a breach of confidentiality that harmed or embarrassed a subject (Schrag, 2010, pp. 63–67; Survey Research Center, 1976, pp. 58, 27). Schrag's *The Institutional Review Blog* describes cases where well-intended IRB regulations were costly and harmful.

7.3.3   No Evidence of Benefit from Regulation of Science

We searched for evidence that regulations on the design of experiments, administered by IRBs and their equivalents in other countries, are more effective at protecting human research subjects than those designed by the experimenters without review board oversight.

Extensive reviews in the books by Schneider (2015) and Schrag (2010), and the report by the Infectious Diseases Society of America (2009) suggest to us that scientific progress is seriously impeded by regulations. We have been unable to find evidence that challenges their findings.

In effect, the US Federal Government now has the ability to control research done at almost all universities. These regulations slow scientific progress and increase costs.

More generally, we are unaware of any scientific research that supports claims of beneficial effects of any government regulation. A seminal work was Winston's (2007) report. His analysis of nine industries before and after they were deregulated in the late-1970s and early 1980s found that customers were better off without the regulation in the case of eight of the industries and that there was little difference in the case of the ninth.

In 2017, we created a website to assess the effects of government regulation in any field, including science. We called the site IronLawofRegulation.com. The Iron Law is most commonly stated as "There is no form of market failure, however egregious, which is not eventually made worse by the political interventions intended to fix it." We opened the site with a challenge to provide scientific evidence that regulations have produced net benefits: "We have, to date, failed to find any experimental studies based on laboratory experiments, field experiments, or natural experiments. We appeal to you for help. The only way to show when to use regulation is to find studies that show the conditions under which it is successful. In other words, we need to find exceptions to what seems to be an iron law." That challenge has gone unanswered.

Our inability to find scientific evidence that supports the hypothesis that regulations provide net benefits over the long run is consistent with Cecil and Griffin's (1985) conclusion from their review of legal policies on US government agencies' data sharing policies that they "base regulatory findings on the conclusions of the research, and yet thwart access to the records by persons and organizations the agency does not wish to have them" and that agencies "structure their relationships with research grantees and contractors in such a way that controversial or sensitive federal research records relied on by the agencies will be beyond public scrutiny" (p. 180).

Milgram's obedience to authority experiments show that an instruction from an authority figure tends to remove responsibility from individual researchers (Milgram, 1974). Given that the researcher who wishes to conduct experiments is likely to know more about the benefits of studying the problem and any risks that might be involved – and how to eliminate or limit the latter – it is hard to see how assigning responsibility and authority to a distant rule maker and a decision-making committee could be expected to improve the situation.

7.3.4   Speech Restriction and Self-Censorship

In the past, governments have often restricted scientists' free speech. Consider the response to Galileo's calculations of the movement of planets, or the Soviet government's endorsement of Lysenko's flawed theories about plant breeding and the ensuing persecution of scientists who disputed Lysenko's theories (Medvedev, 1969; Miller, 1996).

The government's hand does not need to be as heavy as it was with Galileo or in the Soviet Union in order to restrict speech. Speech restriction is a natural consequence of governments' large share, and near monopoly in some cases, of grant funding supported by IRBs and ethics committees that are motivated to keep that funding flowing.

The US Supreme Court was asked to take a case involving commercial free speech. The court declined the case. Justices Thomas and Ginsberg wrote a dissenting opinion, saying that the case was important, and suggested that research be done.

Some years after that case was heard, we were asked to do experimental research on the effect of speech restriction to provide evidence for a court case involving the American Academy of Implant Dentistry. Our experiments on government regulation of speech concluded that it was harmful to consumers and producers. The Florida Court agreed with us (*Ducoin, et al.* v. *Viamonte Ros, et al.*, 2009). In the course of our research, we found 18 experimental studies on commercial speech restrictions, and all of the findings were consistent with ours (Green and Armstrong, 2012).

In addition, a comprehensive review of evidence on government mandated disclosures of all kinds – including those associated with informed consent and IRBs – failed to identify a single mandatory disclosure that provided benefits that were greater than the costs (Ben-Shahar and Schneider, 2014).

Freedom of speech – which includes the freedom to study any problem, and to do so without restrictions – is vital to scientific progress. In the United States, for example, current regulations targeted at organizations that receive government funding can restrict free speech by specifying what can and cannot be studied, how the study must be designed, and how it must be reported.

8 WHAT IT TAKES TO BE A GOOD SCIENTIST

Genius is "1 per cent inspiration and
99 per cent perspiration."
> Thomas Edison, as quoted in *Idaho Statesman* (1901, May 6,
> p. 4, col. 3)

The discovery of useful scientific knowledge depends primarily on the scientist. A scientist's research can be greatly aided by research assistants, advisors, reviewers, journals, and all those fostering a creative environment. Nevertheless, we believe that the primary responsibility for conducting useful scientific research usually falls on individual scientists, sometimes two scientists, but rarely more than two.

In his book, *The Sports Gene*, Epstein (2013) provides convincing evidence that success in sports is based on one's genes, as well as on training. Inherited traits appear to be important for success in science as well.

In this chapter we discuss evidence on characteristics that are common to successful scientists. We also provide a checklist for assessing self-control, a short test on logical thinking, some comments on doctoral programs, and advice to act in a way that is consistent with the scientific method in all roles in which you are identified as a scientist or expert.

8.1 General Mental Ability

Mahoney's (1976, pp. 36–39) review of studies suggested that scientists on average are above the 95th percentile for general mental ability

(GMA), which is equivalent to an IQ score approaching 125. Schmidt (2018) suggested an IQ of 130 as the lower limit for intellectual achievement at a high level. The figure of 130 is the average IQ of US PhDs.

8.2 Family Propensity

Francis Galton's (1874) statistical analysis of survey responses from 100 renowned scientists of his time in England suggests that heredity plays a role in scientific achievement. For example, Galton was Charles Darwin's half-cousin. Among the relations of the 100 scientific men, he found 28 fathers, 36 brothers, 20 grandfathers, and 40 uncles who were notable in some way. He concluded that the scientists' notable ancestors were more or less evenly balanced between the paternal and maternal sides.

If, like Scott and Kesten, you are not related to a famous scientist, do not despair. Many productive researchers have had ancestors who did not practice science, but nevertheless passed on the necessary genes.

Consider also the story of a blacksmith's son apprenticed as a bookbinder with little formal education. He was Michael Faraday. Faraday read books that he was binding, became fascinated by science, and devoted himself to scientific experiments and attending public lectures before he was 20 years old. He became a great scientist and made important discoveries in chemistry, electricity, and magnetism (Williams, 2019).

Galton noted that his list of eminent scientists included "men who have been born in every social grade from the highest order in the peerage down to the factory hand and simple peasant" (1874, p. 21). The fathers of 40 percent of them were merchants, manufacturers, or bankers. While one-third had attended Cambridge or Oxford university – to some extent an indication of inherited talent, expectations, and access in Galton's time – and another third had attended another UK university, a third had not attended any university. Galton was not able to distinguish between the three groups in terms of their scientific achievements. Moreover, while 26 of the eminent scientists praised their formal education, 42 considered that theirs was not useful or worse.

While Galton was convinced that heredity played an important role in success as a scientist, he recognized that "[t]he effects of education and circumstances are so interwoven with those of natural character in determining a man's position among his contemporaries, that I find it impossible to treat them wholly apart. Still less is it possible completely to separate the evidences relating to that portion of a man's nature which is due to heredity, from all the rest" (1874, p. 39). The debate on the relative contributions of nature and nurture continues to this day.

8.3   Early Desire to Do Scientific Research

> ... men of science are not made by much teaching, but rather by awakening their interests, encouraging their pursuits when at home, and leaving them to teach themselves ... throughout life.
>   Francis Galton (1874, p. 257)

Scott asked a friend – one of the best-known scientists in the United States – when he realized that he should be a scientist. He was in early high school, he said, when a new neighbor moved next door. Seeing the man out in the yard, he asked him, "What do you do?" He said that he was a scientist at the local university and that he wrote papers about his studies of economics. Scott's friend's reaction was, "and they pay you for that?!" He had found his calling. Other colleagues have shared similar stories with Scott.

As with Faraday, successful scientists tend to realize that science is their calling at an early age; typically before the age of 12 years (Feist 2006). The desire to improve knowledge and practice is married to the ability to discover solutions. The trait is similar to that found in entrepreneurs, as is perhaps suggested by Galton's (1874) finding that 40 percent of 100 eminent scientists' fathers were merchants, manufacturers, or bankers.

Thomas Edison was born poor and had no advantages. He attended school for only a few months. He learned how to do scientific work by reading, helped by his mother. He was an entrepreneur who started his first business at age 13 and eventually founded 14 companies. He obtained his first patent at age 22. Eventually he had 1,093 patents in a variety of areas such as sound recording, telephony, electricity, and electric lighting, motion pictures, batteries, and chemicals

In 1920, Filo Farnsworth, a 14-year-old Idaho farm boy, had an idea when looking at the pattern when plowing a field. He thought this might be a clue for how to develop an electronic television. At the time, the Radio Corporation of America (RCA) had been trying to invent television. In 1928, he held the first successful demonstration of electronic television, to the chagrin of RCA. Farnsworth went on to make many other useful scientific findings.

Isaac Newton was 24 when he began work on gravitation. Albert Einstein was 26 when he developed his theory of relativity.

8.4   Personality

> ... they and their parents had the habit of doing what they preferred,
> without considering the fashion of the day. The man of science is
> thoroughly independent in character.
>   Francis Galton (1874, p. 124)

Galton (1874) found that eminent scientists were almost universally independent of character, with half of his 100 respondents having the quality "in excess," and only two being below average in that quality. He also found that the eminent scientists were particularly energetic from young to old age. Forty of the scientists described having great energy and endurance in sporting and leisure pursuits as well as in their scientific researches. Only two reported having below average energy.

In their review of research on the productivity of scientists, Rushton et al. (1987) concluded that productive scientists are less social. They are serious, independent, and self-sufficient. Productive scientists also have "radical imaginations." They are *not* concerned about doing better than others. Instead, they are motivated primarily by the intrinsic rewards of working on interesting research with the prospect of making useful discoveries.

A meta-analysis of attempts to determine the difference between the personalities of scientists and non-scientists concluded that scientists tend to be more open to new ideas, independent, and driven. They are also generally more conscientious (Feist, 1998).

Successful scientists have been consumed by the desire to do scientific research, and devoted long hours to their passion. They found important problems and made useful discoveries.

8.5   Motivated by Intrinsic Rewards

> Their independence of spirit ... [is] not conducive to ... competition:
> they doggedly go their own way, and refuse to run races.
>   Francis Galton (1874, p. 258)

Galton observed in his time that those who chose science did so in spite of the profession of scientist being less remunerative and less prestigious than other pursuits (Galton, 1874). Things have changed since then. Since governments became involved in funding research, grant-winning scientists have been paid generous salaries. Yet, extrinsic rewards, *other than positive verbal feedback* – e.g., "Wow, that is really impressive work you did, Charles" – can harm performance.

A meta-analysis of 128 studies included 101 that tested the effect of extrinsic rewards on free-choice behavior (Deci et al., 1999). Only the 21 studies that tested rewards in the form of purely verbal praise showed a positive effect. Of the 92 studies testing tangible rewards, all had a substantially negative effect on performance except for the nine studies that examined *unexpected* rewards, which had no effect on average (Deci et al., 1999, fig. 1).

More recently, Locke and Schattke (2019) have theorized that the intrinsic–extrinsic dichotomy is an oversimplified characterization of motivation type. They argue that intrinsic motivation is best characterized as pleasure obtained from undertaking the activity; the pleasure of performing the task *well* is a different type of motivation that they call achievement motivation. For example, some people enjoy fishing, even when they catch no fish. They also argue that extrinsic motivation has been too narrowly defined and is better defined as doing something that may or may not be enjoyable as a means to the end of pleasure later, as with putting away firewood for a warm house next winter.

Locke and Schattke (2019) propose that an occupation that provides all three types of reward is what people should be looking for. We suggest that you will struggle to succeed *as a scientist* if you do not get intrinsic rewards from your career.

A survey of 333 authors in the *Journal of Consulting and Clinical Psychology* asked about their primary reason for publishing. Sadly, 35 percent of the respondents said that it was part of their job requirements (Kendall and Ford, 1979). We expect that if the same questionnaire were run in 2022, the majority would respond in that way.

8.6   Self-Control

> Steady perseverance is a third quality on which great stress is laid.
> Francis Galton (1874, p. 103)

Galton wrote of his 100 eminent scientists that impulsiveness is was only claimed by five of them, but even among those five there was also a measure of tenacity. Mischel's (2014) "marshmallow study" suggested that self-control, or its absence, is apparent at an early age – 5-year old children were given a marshmallow and were told that they would get another one if they still had the first one when the person who gave it to them returned from a short errand. The children who exhibited self-control went on to be more successful than the others in later life. While self-control seems to be inherited, Mischel (2014, ch. 18) concluded – albeit without experimental evidence – that training led to some success.

There are other ways of assessing your self-control, such as listing examples in your life when you have planned a course of action and stayed with it, or when you deferred actions and planned what you wanted to do.

For a structured approach, consider Tangney et al.'s (2004) "Brief Self-Control Scale" (BSCS). They found that scores on the scale were predictive of students' grade point averages and positively related to desirable behaviors and character traits. They found no evidence that more self-control (higher scores) was in any way detrimental, even for the highest scores. This measure seems to be widely used; at the time of writing, the BSCS had been cited more than 4,000 times according to Google Scholar.

One study used the BSCS questionnaire in combination with other measures of self-control and found that self-discipline was a better predictor of 140 eighth-graders' final grade point averages than were IQ or general mental ability (Duckworth and Seligman, 2005).

We adapted the questionnaire for the BSCS as Checklist 8.1: Self-assessment of self-control. The only change that we made was to the answer format.

Complete the checklist for yourself and compute your score. The score is not for sharing, but rather for helping you assess whether you could become a productive scientist.

The mean score of the 606 undergraduate psychology students who completed the BSCS questionnaire was about 40, and the standard deviation was about 8.5. Given the importance of self-control for scientists relative to the general population, you might want to know that you are in the top, say, 20 percent of a group equivalent to the one in the study. Assuming a normal distribution of scores, that means that you want a score of 47 or better to conclude that scientific research might be a good career option for you.

Checklist 8.1 Self-assessment of self-control

Indicate how much each of the following statements reflects how you typically are (not at all to very much) by circling the appropriate number in one of the five columns to the right.*	Frequency rarely . . . often				
I am good at resisting temptation	1	2	3	4	5
I have a hard time breaking bad habits	5	4	3	2	1
I am lazy	5	4	3	2	1
I say inappropriate things	5	4	3	2	1
I do certain things that are bad for me, if they are fun	5	4	3	2	1
I refuse things that are bad for me	1	2	3	4	5
I wish I had more self-discipline	5	4	3	2	1
People would say that I have iron self-discipline	1	2	3	4	5
Pleasure and fun sometimes keep me from getting work done	5	4	3	2	1
I have trouble concentrating	5	4	3	2	1
I am able to work effectively toward long-term goals	1	2	3	4	5
Sometimes I can't stop myself from doing something, even if I know it is wrong	5	4	3	2	1
I often act without thinking through all the alternatives	5	4	3	2	1
Sum the circled figures to calculate a total score between 13 and 65	TOTAL SCORE [ ]				

*Adapted from Tangney et al.'s (2004) "Brief Self-Control Scale."

The following self-assessment test might provide additional insight to you if you are considering a career as a scientist (cover the answers in the box while you complete it):

1. A bat and a ball cost $1.10 in total. The bat costs $1.00 more than the ball. How much does the ball cost? _____ *cents*
2. If it takes 5 machines 5 minutes to make 5 widgets, how long would it take 100 machines to make 100 widgets? _____ *minutes*
3. In a lake, there is a patch of lily pads. Every day, the patch doubles in size. If it takes 48 days for the patch to cover the entire lake, how long would it take for the patch to cover half of the lake? _____ *days*

Answers to Self-Assessment Questions

The test, called the Cognitive Reflection Test, was developed by Frederick (2005). It is easy to get the wrong answer. Yet if you pause and reflect on the problem, especially by writing down how you arrived at your answer, you are likely to get the correct answer.

The correct answers are 5 cents, 5 minutes, and 47 days.

When the problem was presented to convenience samples of students at 10 universities, the differences were striking: Michigan State and the University of Toledo were at the bottom – average number of correct answers 0.79 and 0.57 respectively, out of 3 – while Massachusetts Institute of Technology (MIT) and Princeton were at the top with 2.18 and 1.63.

The test has been used by many organizations for hiring decisions.

8.7   Skepticism

Skepticism is critical for those bent on making important and useful findings. The *Oxford English Dictionary* describes a skeptic as "a seeker after truth who has not yet arrived at definite convictions."

Skeptics show little deference to authority or to tradition. Mahoney concluded that "The [scientist] who takes an unconventional path is often faced with isolation, and occasionally with persecution" (1976, p. xiv).

How do people respond when a researcher uses science to address an important problem and the results challenge current understanding? Barber (1961) gave examples of famous scientists who met resistance from other scientists when they worked on important problems that challenged common beliefs of their time. For example, Galton (1872) did research on the effectiveness of prayer and concluded that it was not effective.

Imagine that you are participating in the following experiment. You are one of a group of seven shown the image of a "standard" line on one card and three lines on another card. You are each in turn asked to say which of the three lines was the same length as the standard line. The standard-length line seems obvious to you, but the other six people in your group are asked to respond before you are, and they all picked another line. How would you respond? That question was addressed in an experiment by Asch (1955). The 123 research subjects went along with the majority's incorrect response for about 37 percent of their standard line picking judgments.

Do you find it easy to disagree with groups of people? *Can you think of situations where you stood your ground even though almost all people disagreed with you?* Bear in mind that skepticism is antisocial: people are happier when they associate with people who agree with them. Recently, it has become common to label people who are skeptical of *popular beliefs* as "deniers." In earlier times, the term was "heretics."

How would *you* respond if you were presented with scientific evidence that challenged one of your strongly held beliefs? That has happened to us on a number of issues and we love it when people provide evidence showing that we were wrong. For example, when Scott was a student at Carnegie Mellon, he enjoyed discussing ideas about economics with his suitemate, Ed Prescott (2004 Nobel Prize in Economics). We agreed on nearly everything. For example, we agreed that some government regulations were necessary. A neighboring student in the dorm occasionally took part in our discussions. He disagreed with us on regulation, and he had scientific evidence for his position. Scott revised his beliefs beliefs when he learned about the experimental evidence.

Also consider Sinclair Lewis's novel, *Arrowsmith* – in which a medical researcher used experiments to challenge the government's position on how to stop an epidemic (Lewis, 1925). It provides a fictional account of the obstacles faced by skeptical scientists.

In the preface to his book, *The Scientist as a Rebel*, Freeman Dyson states that "Benjamin Franklin combined better than anyone else the qualities of a great scientist and a great rebel. As a scientist, without formal education or inherited wealth, he beat the learned aristocrats of Europe at their own game" (Dyson, 2008, p. ix).

Consider reading biographies of skeptical scientists. Scott's favorites are Stanley Milgram's biography, *The Man Who Shocked the World* (Blass, 2009), and Julian Simon's autobiography, *A Life Against the Grain* (Simon, 2002). Kesten is inspired by the story of a blacksmith's son and bookbinding apprentice who became a scientific great – Michael Faraday (Williams, 2019).

8.8   Your Decision

A career in science can be appealing. For example, a study of 291 famous men in science, music, politics, philosophy, arts, and literature found that "with few exceptions, these men were emotionally warm, with a gift for friendship and sociability" (Post, 1994, p. 22) The 45 scientists in particular had more stable marriages and were judged to have much lower levels of psychiatric disorders than the other famous men.

Cole (1979) presented statistics on the age and productivity of scientists in chemistry, geology, mathematics, physics, psychology, and sociology. They ranged in age group from "under 35" to "60-plus." He found little difference in productivity by age group, whether measured by the number of papers or the numbers of citations their papers received. Consistent with Cole, the scientists that we know have maintained their skills over time, and their desire to keep working persists.

If you struggle to imagine anything else as satisfying as research – doing proper experiments to make useful discoveries – then science is likely the right profession for you.

If you are not sure that you are destined to be a scientist, we expect that getting a PhD and a position at a university will do little to improve your ability to make useful scientific discoveries or to improve your happiness. If you have not experienced that calling, you should be wary about being a scientist.

If you are still equivocating, consider the following statistics. A survey was conducted on students in eight PhD programs in economics: Columbia University's, Harvard's, Michigan's, MIT's, Princeton's, UC Berkeley's, UC San Diego's, and Yale's. It found that:

- 18 percent of the students experience moderate or severe depression;
- 11 percent think about suicide in a two-week period;
- 26 percent think their work is useful always or most of the time, compared with 63 percent in the working age population in the United States (Barreira et al., 2018).

If you have decided you do want to be a scientist, consider where you are likely to find the best environment for your research: Private corporations? Independent work? Universities, and in which countries? Think tanks? Wealthy donors? If you are interested in universities in the United States, you might want to ensure that you can have freedom to conduct the research you want to do given that almost all US universities facilitate government control over research.

8.9   Navigating a Doctoral Program

Our advice for doctoral students is to devote yourself to doing useful research. Take courses only if they are useful in advancing your ability to do scientific research. Scott's recollection of his time as a PhD student at MIT was that no courses were required. The dean at MIT's Sloan School of Management advised PhD candidates to finish in three years and many did so. Similarly, no courses were required or expected when Kesten did his PhD by thesis at Victoria University of Wellington in New Zealand. Students were encouraged to write research papers and to present them at conferences.

Find a mentor who could help you develop as a scientist. Your supervisor might perform that role, but you should also aim to be in contact with the leading researchers in your topic area.

8.10   Acting as a Scientist in all Relevant Roles

Scientists are often asked by the media for newsworthy comments. Our policy is to avoid commenting on any topic on which we lack relevant scientific knowledge – in other words, avoid opinions.

Scientists are sometimes offered money to provide expert analysis and advice – such as in consulting arrangements – and testimony – such as for a court of law. If, after examining a brief, you decide that the position of the party requesting your services may be contrary to the evidence, we recommend declining the request. If you do agree to a consulting arrangement or to provide testimony, document your commitment to objectivity.

We have stated the following conditions when testifying as expert witnesses in court cases. Similar conditions could be required when engaged as a consultant.

"My objectivity and integrity are central to my testimony. I will only act as an expert witness under the following conditions:

1. My testimony addresses matters on which I am familiar with the scientific evidence.
2. I will be fully informed by the lawyers about relevant matters in the case.

3. The case deals with an important issue that will affect others in a positive way. In short, there must be some wider benefit.
4. To help ensure that my testimony is free of bias, I only work on an hourly basis, with no contingency fee.
5. To help ensure that my testimony will not be thought to be biased, I must be fully paid (including expected time for testimony) prior to testifying before a Court or Arbitration Board.
6. I will abide by the Guidelines for Science.

The sixth condition was added in 2019.

9 HOW SCIENTISTS CAN DISCOVER USEFUL KNOWLEDGE

We believe that the primary role of a scientist is to make discoveries that can help to improve peoples' lives, whether directly such as through the discovery of a vaccine against a disease or indirectly such as through the invention of a procedure that can improve efficiency or lead to better decisions. In this chapter, we provide practical advice on how to identify important problems, how to choose which important problems to research, how to design a study, how to collect data, and how to analyze them.

9.1 Identifying Important Problems

The first step to discovering useful scientific knowledge is to identify problems. In this section, we describe six steps for generating creative ideas for problems to study and possible solutions, and methods for evaluating your ideas. The steps are summarized in Checklist 9.1, Identifying important problems.

When a problem jumps out at you, such as with Milgram's idea to study blind obedience, you can omit steps 1–4. Many people were intrigued by the question of to what extent subjects would blindly obey an authority when instructed to hurt another person when acting as a research assistant conducting an experiment.

When Scott started to plan his PhD research, one of his professors advised him to seek an important problem in a field that was not already being studied by many others. That turned out to be good advice for him.

Checklist 9.1 Identifying important problems

1. Work independently	☐
2. Problem-storm using brainwriting	☐
3. Develop solutions alone, ignoring others' solutions at first	☐
4. Get close to the problem to learn about current solutions	☐
5. Seek help from others	☐
6. Build on potential solutions while avoiding evaluation	☐

Do not rush to judgment on selecting a problem. Selecting a problem is the most important decision you will make, and it requires much creativity. How to choose *which* important problem to study is the subject of Section 9.2 of this chapter.

Given that the brain is the organ that uses the most energy in our bodies, it seems to make sense to take rest breaks. Early in his career, Scott found that taking short naps during the day contributed to his creativity. In total, his sleep hours were about the same as before adopting the strategy. He is able to fall asleep in minutes, which is helpful.

Adult napping is a fairly common behavior. While waking from a deep nap may be disorienting, there is little evidence that it is harmful to productivity. Overall, nappers reported less sleepiness and fatigue, and better mood and performance after taking a nap (Dinges, 1992).

A review of the literature on the effects of napping on cognitive function found that naps from five minutes to two hours "have been shown to have some benefits to cognition" (Lovato and Lack 2010, p. 157). Naps were more effective than caffeine and stimulant medications at improving mood, alertness, and performance.

Scott was a runner until he was 75. He managed to convince himself that running and other types of exercise improved his creativity. Scott is not alone in finding benefit from exercise. One study tested the value of a brisk walk. In four experiments that tested the effect of exercise under different conditions – inside on treadmill or outside – compared to sitting, found walking roughly doubled the number of creative ideas that the 176 subjects generated (Oppezzo and Schwartz, 2014).

Scott has additional guidelines for creativity but found no evidence to support them. They include: (a) when working on a paper or a book, spend time on it every day if possible; (b) work on a difficult section just before going to bed so as to allow your subconscious to keep working on it (Smith, 2005, describes anecdotes and some evidence on the role of the subconscious in finding solutions); (c) start the next day on that difficult section.

9.1.1 Work Independently

Identifying important problems requires creativity, along with an awareness of your own interests and skills. You are the only one who knows those; so, do not bother asking others to suggest what you should work on.

We can easily think of individuals who have made useful scientific advances, and it is possible to recall groups of two, but it is difficult for us to think of groups of three or more people who have done so.

Social loafing almost always occurs within groups that are working toward a pooled outcome, especially when the loafers expect co-workers to do a good job. That was the conclusion of a meta-analysis of 78 studies. The study also found that the larger the group, the greater the loafing (Karau and Williams, 1993).

If you do work with others, avoid having face-to-face meetings as much as is feasible. Doing so is likely to lead to better decisions (Armstrong, 2006).

9.1.2 Problem-Storm Using Brainwriting

The wording of a problem can limit the search for solutions. Thus the need for problem storming: state the problem in different ways, then search for possible solutions for each statement of the problem. For example, politicians who are concerned that higher education should be more effective usually ask, "How can we improve teaching?" An alternative question is, "how can we improve learning?" The latter framing of the problem yields recommendations that differ sharply from those of the first.

Brainstorming was proposed in 1940 for increasing creativity. Experiments have shown that, when used as directed, it produced more creative ideas. Unfortunately, it is an expensive procedure, as it requires the use of a trained facilitator along with scheduling group meetings. As a result, the method is rarely used properly in practice. In addition, while brainstorming is useful in that it does not allow negative evaluations of ideas, it does allow for supportive statements, and that leads groups to prematurely conclude that the creativity task has been completed.

Brainwriting is much simpler and less expensive than properly conducted brainstorming. It involves setting a goal for the number of creative ideas to suggest, then independently writing all of the ideas you can think of, avoiding evaluating the ideas until you finish compiling your list.

A review of the evidence on the various forms of brainstorming concluded that avoiding interactions while ideas are generated by employing some form of nominal group – such as brainwriting or electronic brainstorming – is more productive of ideas than standard brainstorming (Kerr and Tindale, 2004).

9.1.3   Develop Solutions Alone, Ignoring Current Solutions at First

We have no experimental evidence on the issue, but inventors, including Sir James Dyson, have described working on problems without considering current solutions (Gelles, 2018).

9.1.4   Get Close to the Problem to Learn About Current Solutions

One of the nice things about engineers is that their job requires that they get close to the problem and observe what is happening. This is not required for academic research. Many academics look only for convenient data. For example, a convenience store chain wanted to improve sales by having their salespeople smile more often. An analysis of the data found that sales were lower when the clerks smiled more often. In that case, however, the researchers also visited the stores, talked to salespeople, and worked in the store themselves. They observed that when many customers were in the store, both sales personnel and customers changed their behavioral norms to cope with the increased stress, with the consequence that there were fewer pleasantries such as smiling (Sutton and Rafaeli, 1988).

Current approaches should stand as reasonable hypotheses. For example, when engineers identify a problem, they typically search for multiple reasonable solutions, and then conduct experiments to test them.

9.1.5   Seek Help from Others

When you are ready, share the design of your experiment with others. Ask them to predict the findings. If their predictions are wrong, you might be onto something important. For example, Kesten showed that the findings from his studies on the predictive validity of alternative methods for forecasting decisions in conflict situations differed from expectations. In particular, 77 attendees at talks on Kesten's research expected 50 percent of game theorists' forecasts to be accurate for eight conflicts when guessing could be expected to provide 28 percent accurate forecasts.

Moreover, they expected 40 percent of novices' simulated interactions – a form of role-playing – to provide accurate forecasts. In contrast, 31 percent of the 101 forecasts by game theorists and 62 percent of the 105 forecasts from simulated interaction were accurate (Green, 2005).

Do not ask people if they are surprised after they see the findings. Three experiments showed that people seldom express surprise, no matter what the findings (Slovic and Fischhoff, 1977).

Another useful approach is to write a "press release" describing your intended study and any expected findings. Ask for feedback on the methods and findings. Do people think your study is important?

Hal Arkes, who has made important discoveries in the management sciences, told us about his "Aunt Mary Test" for evaluating new ideas. At Thanksgiving each year, his Aunt Mary would ask him to tell her about his important new research. When she was skeptical about a research idea, he said, "I didn't always abandon it, but I always reevaluated it, usually resulting in some kind of modification of the idea to make it simpler or more practical" (Hal Arkes, personal communication, September 9, 2015).

With Hal Arkes's Aunt Mary in mind, ask a number of people with common sense – but who are not necessarily experts on the specific problem – whether a problem that you are considering is important. Doing so will require that you frame your proposed research problem in plain words.

While we urge scientists to avoid face-to-face meetings, there are situations that require meetings, such as when disagreements arise about the design of an experiment, or when developing standards for a given project. When that happens, consider the use of the book, *Problem-Solving Discussions and Conferences* (Maier, 1963). It describes the findings of creative experiments on ways to increase the effectiveness of group meetings. Scott regards it as the "bible on running group meetings."

9.1.6 Build on Potential Solutions While Avoiding Evaluation

> If we watch ourselves honestly we shall often find that we have begun to argue against a new idea even before it has been completely stated.
> *Wilfred Trotter* (1941, p. 186)

It natural for people to think first of what is bad rather than what is good about a new idea. When considering a new idea, think of ways to improve it *prior to any evaluation.* That approach has been called "building" on each idea, and is especially important when asking for ideas from others.

Once you have an idea for a solution, assume that you have been prohibited from using that solution. Then develop another solution. The technique was developed by Maier and Hoffman (1960) and is known as the second-solution technique. Evaluation of possible solutions can then be made after each has been presented in its best light.

We now describe guidelines for conducting a useful scientific study under four broad headings:

1. Selecting a problem
2. Designing a study
3. Collecting data
4. Analyzing data

For each step, we provide guidelines for complying with the scientific method. The guidelines rely heavily on the literature. The guidelines are summarized in the Conducting a Useful Scientific Study checklist (Checklist 9.2) overleaf.

9.2 Selecting a Problem

Checklist 9.1 provides guidelines on how to identify important problems. In this section we describe how to choose among them.

Checklist 9.2 Conducting a useful scientific study

Selecting an important problem

1. Choose a problem for which findings are likely to provide benefits without duress or deceit	☐
2. Be skeptical about findings, theories, policies, methods, and data when lacking experimental evidence	☐
3. Consider conducting replications and extensions of papers that address important problems	☐
4. Ensure that you can address the problem impartially	☐
5. If you need funding, ensure that you will nevertheless have control over your study	☐

Designing a study

6. Summarize existing scientific knowledge about the problem	☐
7. Develop multiple reasonable hypotheses with specified conditions prior to any analysis	☐
8. Design a study that minimizes the risk of harm to subjects	☐
9. Pretest experiments	☐
10. Warn subjects that they might find the task unpleasant	☐
11. Consider role-play as alternative to experimentation	☐
12. Design experiments that estimate effect directions and sizes so as to identify the best hypothesis	☐

Collecting data

13. Obtain all valid data	☐
14. Ensure that the data are reliable	☐

Analyzing data

15. Use models that incorporate cumulative knowledge	☐
16. Use simple methods	☐
17. Use multiple validated methods	☐
18. Draw conclusions only on practical importance of effects	☐

9.2.1 Choose a Problem for Which Findings Are Likely to Provide Benefits without Duress or Deceit

Research can produce useful findings only if the problem is important. An important problem is one for which new scientific findings could benefit people without harming others. Examples include developing more efficient, effective, and safer products or services; reducing waste, crime, and conflict; improving health or life satisfaction; protecting individual freedoms; and discovering cost-effective ways to ensure equal opportunity under the law.

9.2.2 Be Skeptical about Findings, Theories, Policies, Methods, and Data When Lacking Experimental Evidence

Science is more likely to be useful if it addresses problems that have been the subject of few, if any, proper *experimental* studies. There are many important problems that could benefit from experimental evidence. For example, proponents assert that game theory can be used to make better decisions in situations that involve conflict or competition between parties. Unable to find experimental evidence to support that claim, Kesten conducted experiments – alluded to above – to test the predictive validity of game theory. He found that game theorists' predictions of what decisions would be made in real conflict situations – including union-management disputes, a civil protest, a hostile corporate takeover attempt, and a military-diplomatic standoff – were no more accurate than unaided guesses by naive subjects (Green, 2002, 2005).

Semmelweis's experiments provide a classic example of a researcher taking a skeptical approach to his contemporaries' untested practices. He found that when medical doctors delivered babies, more mothers died from puerperal fever than when midwives handled the deliveries. He speculated that the doctors might be transmitting the disease on their hands from dissecting cadavers before visiting the maternity ward. When he did an experiment in which the doctors were required to wash their hands in a solution of chlorine, he found that deaths among expectant mothers fell from 14 percent in 1846 to 1.3 percent in 1848 (Routh, 1849). Despite the strength of his findings, doctors resisted making any changes in practice.

More recent examples include the standard treatment for stomach ulcers and the treatment of head injuries with anti-inflammatory drugs. Neither had been validated.

In the case of stomach ulcers, the treatment was to live a life avoiding stress, spicy foods, and alcohol in order to alleviate the symptoms. When two researchers in Western Australia proposed that the cause was a particular bacterium, and that the condition could be cured by a dose of the appropriate antibiotic, they were ridiculed, had their papers rejected, and careers held back. Their evidence was eventually accepted, and stomach ulcers are now easily cured (Nobel Media AB, 2005).

In the case of head injuries, when anti-inflammatory drugs were developed, they were applied to head injury patients in the belief that they would help by reducing brain swelling. Years later a large experiment was conducted to test that hypothesis. It was terminated early when it became clear that proportionately more of the patients were dying when they were given the drug that had been the standard treatment for many years (CRASH, 2004).

9.2.3   Consider Conducting Replications and Extensions of Papers That Address Important Problems

Replications and extensions or variations of *useful scientific studies* are important regardless of whether they support or contradict the original study. Replications of useless or non-scientific studies, on the other hand, have no value. Direct replications are useful when there are reasons to be skeptical about findings relating to an important problem. Extensions (variations) of important scientific studies are useful, as they can provide evidence about the conditions under which the findings apply.

If you believe that you have identified cheating in a published paper that addresses an important problem, the appropriate response is to first let the authors know that you believe you have found an error. If you do not get a satisfactory response, contact the editor of the journal in which the paper was published and advise them of your concern. Consider undertaking a replication of the study, and ask the editor whether the journal would publish it.

9.2.4   Ensure That You Can Address the Problem Impartially

Once you have a list of important problems, examine each to see if you could develop experiments to test leading reasonable alternative hypotheses. Can you structure an experiment such that your favored hypothesis might not dominate? If not, find another problem.

Alternatively, work with a co-author with opposing views on the problem to design a study that you can both agree would be a fair test. That solution was described to us by Joel Kupfersmid in a personal communication in February 2019. He told us that the professors on his dissertation committee had opposite views and, as a result, his study was designed in a way that met with the approval of both sides of the argument.

9.2.5 If You Need Funding, Ensure That You Will Nevertheless Have Control Over Your Study

Do you really need funding? Think about how you can do the research in a more economical manner. Some researchers use their own money to cover the costs of their research, which helps to motivate the search for cost-effective approaches. Some universities provide research budgets for faculty to be spent as the researchers see fit. Amazon's Mechanical Turk can provide useful resources cheaply for some studies. With some creativity, much useful research can be done on a modest budget.

Researchers must retain control over all aspects of the design, analysis, and writing to ensure their study follows the scientific method. That can be a problem for researchers at universities that receive government funding in the United States and elsewhere. Even if you do *not* receive direct external funding, you may be subject to restrictions by your institution on the topics you are permitted to study, how you must design your experiments, and what you must say in your paper. If you are required to make changes that will harm the design of the study or cause harm to a subject, it is your responsibility to reject those changes.

9.3 Designing a Study

The next seven guidelines describe how to design experiments to ensure objective and useful findings.

9.3.1 Summarize Existing Scientific Knowledge about the Problem

To contribute to useful scientific knowledge, researchers must become knowledgeable about the relevant scientific findings to date. That injunction applies to identifying and using valid research methods as well as to domain knowledge. Literature reviews are expensive and time-consuming – but critical.

In recent years, older studies have been increasingly ignored. The practice of scientific research has changed enormously over the last century. In the course of writing one book on persuasion principles and two on forecasting principles, Scott came to the conclusion that the quality of the older papers was, on average much better than the research that is currently being published in the leading journals.

Ask leading scientists in the area of investigation to suggest relevant experimental papers. Use the references in those papers to find additional papers, as well as citations to those papers.

We believe the search for evidence would be greatly simplified if it were restricted to papers that conform to the scientific method. We do that in our own reviews as it takes only a few minutes to examine the abstract and conclusions to see whether a paper with a promising title complies with the scientific method.

Internet searches should also be used to find relevant papers. However, given the large number of academic works available online, many papers that *seem* promising based on their title and key words fail to provide useful scientific findings. In addition, few papers provide a summary of the findings in the abstract. Moreover, online searches miss many relevant papers, in part because terminology varies across fields. For example, a key-word computer search for studies on forecasting methods in the *Social Science Citation Index* identified only one-sixth of the papers that were eventually identified and cited in Armstrong and Pagell (2003).

To improve internet searches, we suggest that researchers and journals provide information about a paper's compliance with science. Even without that, however, one can quickly use Checklist 3.1, *Compliance With Science Checklist*, for key items such as whether a paper uses experiments to test multiple reasonable hypotheses by looking at the abstract, tables, and conclusions. By quickly, we mean within minutes.

To help ensure objectivity in literature reviews, one should use meta-analyses. The purpose of meta-analyses is to ensure that the review provides a summary of *all* relevant scientific findings. The criteria for inclusion of a paper should be specified before beginning the literature search.

Given the use of pre-specified criteria, meta-analytic reviews are less likely than traditional reviews to omit papers that conflict with the author's favored hypothesis, as was shown by Beaman's experiment

(1991). Beaman sampled reviews from 1981 to 1983 and from 1987 to 1988 for his comparison.

Practical advice on conducting meta-analyses is provided in Schmidt and Hunter's (2015) book *Methods of Meta-Analysis: Correcting Error and Bias in Research Findings.*

Once prior knowledge is obtained it must be summarized. In the mid-1900s, econometricians were expected to undertake a priori analyses to develop models that were consistent with prior knowledge. Doing so was critical for showing how the paper contributed to scientific knowledge. Moreover, it was much more expensive to do regression analysis in the mid-1900s as Friedman described in his appendix to Friedman and Schwartz (1991).

A priori analysis can be used to develop a model prior to undertaking any statistical analysis. That calls for a complete model, with variables and coefficients, based only on prior knowledge. For example, Scott did his PhD thesis on forecasting international sales of cameras. He spent months on the a priori analysis to fully specify a model. He handed his thesis advisor a sealed envelope with the model. The a priori model provided accurate forecasts of sales. When the coefficients were averaged with those from a regression model, the accuracy of the forecasts was further improved (Armstrong, 1985, pp. 191–247).

There are other ways of ensuring that the research builds on prior knowledge. One is to register the hypotheses, as some have suggested. That approach might encourage researchers to advocate for a hypothesis, however. Also, as scientists work on research projects, they might encounter new information about conditions and develop additional hypotheses. If they do, they should explain how and why their hypotheses changed. Another approach is to retain notes and drafts of the research paper. We favor the latter approach as being less likely to lead to advocacy, but PLOS's facility to preregister a research question and a study design and subject them to constructive reviews might address our concerns (see www.plos.org/preregistration).

9.3.2 Develop Multiple Reasonable Hypotheses With Specified Conditions Prior to Any Analysis

Identify current approaches to the problem, your own ideas, and alternative solutions proposed by others. Ask experts in the field to suggest additional solutions. Seek out people who have diverse and relevant knowledge that differs from your own. Think also of how similar problems were solved. Keep in mind that major breakthroughs were achieved by the researchers who dared to test hypotheses that challenged the established beliefs such as on hand washing (Semmelweis), stomach ulcers (Marshall and Warren), and head injuries (CRASH Trial collaborators) as we described above.

When you have promising hypotheses for an important problem, list reasons why each hypothesis might be wrong. Doing so helps to specify the conditions under which the hypotheses are likely to apply.

Consider that your findings might be inadvertently affected by an unobserved factor and thereby result in erroneous conclusions. In the early 1900s, a horse, "Clever Hans," could apparently answer an amazing variety of questions by tapping its hoof. For example, "how much is 3 plus 4?" It was hard for people to believe that a horse could be so clever, so a commission was appointed to find out how he was able to answer. He could answer questions from strangers even when his trainer was out of sight. The investigation stumbled about for some time until it was discovered that the answer lay not in Clever Hans, but in the person who asked the question. Clever Hans was able to answer a question only when the person asking it knew the answer. Without realizing it, questioners were giving cues about the correct answers. When the correct answer approached, the questioner would often show an almost imperceptible head movement, which he was not aware of himself. That was the cue that Hans took to stop tapping (see *Clever Hans* in Wikipedia).

Researchers should test all reasonable hypotheses. Doing so is an essential element of the scientific method. Be warned, however, that if the findings support an unpopular or controversial hypothesis, publication of the paper may be more difficult. When Scott submitted what he regarded as his most important papers with findings from multiple reasonable hypothesis tests, all suffered initial rejection. He would then make changes demanded by journal reviewers that he agreed with, appeal to the editor, and usually face another rejection. That cycle would continue until an editor at some journal would finally overrule the reviewers and publish the paper. In the end, all but one of what he believes to be his most important papers were published.

Which hypotheses should be included among the reasonable? Keep in mind Peirce's doctrine of economy:

> The doctrine of economy, in general, treats of the relations between utility and cost. That branch of it which relates to research considers the relations between the utility and the cost of diminishing the probable error of our knowledge. Its main problem is, how, with a given expenditure of money, time, and energy, to obtain the most valuable addition to our knowledge. Charles Sanders Peirce (1958, para. 140)

Note that the doctrine of economy also applies to selecting research problems ("Identifying Important Problems," Section 1 in this chapter).

9.3.3   Design a Study That Minimizes the Risk of Harm to Subjects

The researcher is best able to design an experiment that is safe for human subjects. Researchers are also best placed to design a consent form for subjects, if one is needed.

Milgram's (1963) experiments on obedience to authority found that the risk of harm to subjects was increased if someone other than the researcher controlled the research procedure. The problem is likely to be greatest when government officials or their agents – such as IRBs and ethics committees – control the design of research, because they have the power to enforce compliance and punish non-compliance. Moreover, without control of the research design, the researcher loses personal responsibility for it.

Researchers should ensure that any paper that goes out in their name complies with their standards for the ethical treatment of subjects. Milgram's (1974, app 1, pp. 193–202), for example, provides a detailed description of the steps he took to ensure subjects in his experiments were treated with respect. In the studies described in Milgram's book, subjects acting in the role of assistants in an experiment were told that they were administering electrical shocks to another person by obeying instructions from an authority figure.

9.3.4   Pretest Experiments

When designing "laboratory" or field experiments that involve human subjects, experimental materials should be tested to ensure that they are fit for purpose and designed in a way that will not be unnecessarily upsetting to the subjects. For example, Milgram considered and used many designs for the obedience experiments, as described in Blass (2009).

Natural experiments and experiments that do not use human subjects also benefit from pretesting designs. In the case of natural experiments, however, shortage of data may constrain pretesting.

9.3.5 Warn Subjects That They Might Find the Task Unpleasant

Describe the risks associated with the experiment, if any. Experiments that use human subjects by their nature involve some deception; at a minimum, not telling subjects the hypotheses, or whether they are allocated to the placebo treatment. Another approach is to pay research participants for their time on the understanding that they will perform a task for the researcher.

Debriefing subjects can help to ensure that they regard the experience of having participated in an experiment as a positive one as, for example, Milgram's (1974) debriefing and follow-up survey of all 856 subjects in his experiments obtained a 92 percent response rate. Only 1.3 percent of them were sorry that they had participated in the experiment. Forty-four percent were "very glad" and 40 percent were "glad" that they had participated.

9.3.6 Consider Role-Playing as an Alternative to Experimentation

Some situations make experimentation with human subjects difficult. For example, there might be concern about harm to subjects. (Note that by harm, we do not mean the possibility that subjects might feel uncomfortable: experiments on important problems will likely upset some subjects.) As an alternative to experiments, one might ask participants to play specified roles in a situation that you describe to them.

Motivated by Milgram's blind obedience studies, there were a number of experiments that compared decisions by subjects using role-playing with those from the actual experiments. The first of the attempts to replicate Milgram (1963) without deceiving subjects involved 38 students. The results were remarkably similar to Milgram's, with a majority of the subjects (74 percent) administering the strongest ("extremely painful") shock, compared to Milgram's 65 percent and, in both cases, all subjects "administered" at least 300 volts (O'Leary et al., 1970).

Mixon (1972) used active role-playing in an attempt to replicate Milgram's experiment. One subject pretended to be the experimenter and the other pretended to be the learner. In the actual experiment, 65 percent of subjects were completely obedient and the average shock the subjects administered was 405 volts out of a maximum of 450 volts. Of Mixon's 30 naive role players, 80 percent were fully obedient, and the average shock level was 421 volts.

Further support was provided by Berscheid et al., (1973) in their study that asked subjects to imagine themselves in the role of a typical subject in several experiments that were described to them, and to rate how willing they would have been to participate. Their responses were consistent with the responses of subjects who had participated in replications of Milgram's experiment by Ring et al. (1970) and were asked afterwards whether they would have chosen to participate.

The practical usefulness of role-playing experiments was dramatically demonstrated by the war games played by the staff of the Western Approaches Tactical Unit (WATU) – mostly young women recruited straight out of high school – during WWII. Their role was to develop tactics to combat the deadly German "wolf pack" U-boat attacks on Allied trans-Atlantic convoys, and to teach those tactics to naval officers. Using reports of encounters with the U-boats and crude mock-ups of the encounters on a basement floor, the WATU staff played the roles of the U-boat commanders. In those roles, they were able to recreate enemy tactics, and to then develop tactics that the Royal Navy officers were able to use to counter the threat. As a result, the threat that the U-boats posed to Allied shipping was greatly diminished (Strong, 2017).

Role-playing has been successfully used in other experiments, and has been found to have superior predictive validity for decisions in real conflict situations to predictions from novices' unaided judgments (Armstrong's 2001c); game theorists' predictions (Green, 2002, 2005); experts' unaided judgements (Green 2002, 2005; Green and Armstrong 2007a); and from a structured analysis of expert's analogies (Green and Armstrong, 2007b).

Logic, and Mixon's (1972) research on Milgram's experiments, suggest that role-playing might be *more* predictive of behavior out of the laboratory, especially for situations that would require experimental subjects to believe that they are causing actual harm. The challenge of deceiving subjects to the extent that they *entirely* overcome their prior beliefs that scientists conducting experiments in, for example, the United States and Australia would make sure that no one was harmed is perhaps insurmountable. Moreover, subjects who suspect that they may be being deceived are unlikely to behave as they would outside of the classroom or laboratory. For example, reviews of evidence on whether research participants who reported that they were suspicious that they had been deceived and those who were told they would be deceived concluded that they tended to behave differently to non-suspicious participants in experiments on conformity (Hertwig and Ortmann, 2008).

9.3.7   Design Experiments That Estimate the Effect of Directions and Sizes So as to Identify the Best Hypothesis

When we were very young and could not speak, we needed to experiment to learn. If we did not experiment, our parents would fear that there was something wrong with us.

Natural experiments have obvious validity in that they apply directly to the situation. Lack of control over causal variables can, however, limit their usefulness. Consider the debate about which is more important to one's health: lifestyle or health care. When Russia abruptly ended its support of Cuba's economy, an economic crisis began in 1989 and lasted until 2000. There was a lack of food, health care, and transportation. People had to leave their desk jobs to work in the fields to grow food. An analysis of Cuba's national statistics from 1980 through 2005 found that food consumption decreased by 36 percent and obesity decreased from 14 percent to 7 percent of the population. By 1997–2002, deaths due to diabetes dropped by half, and those due to heart disease by one-third (Franco et al., 2007). That natural experiment had a large sample size and large effect sizes over more than a decade, but it involved only a single country, and it is possible that there may have been other important causal factors at play.

Some problems benefit from the availability of data from many natural experiments, thereby improving reliability and generalizability. Consider gun control: Regulations vary by states, cities, localities, and institutions in the United States, thereby providing large numbers of natural experiments to assess alternative policies. John Lott has been analyzing such findings for over two decades. His conclusions from the analyses of natural experiments have differed substantially from those who rely only on expert opinions or on analyses of non-experimental data (see, e.g., Lott, 2010).

More broadly, the study and analysis of natural experiments that occur as a result of the different social and economic institutional arrangements that evolve in different communities have provided useful evidence against which to test hypotheses. For example, biologist Garrett Hardin (1968) speculated that the future would be bleak because people would exploit natural resources to destruction, an idea that he popularized as "the tragedy of the commons."

Nobel Prize winning economist, Elinor Ostrom, in her (1990) book *Governing the Commons*, provided specific disconfirming evidence. She found that destruction of resources was a far from inevitable outcome for the common pool resource situations – such as fisheries or water for irrigation – that Hardin's hypothesis was concerned with. Ostrom found that people with access to a resource commonly find ways to manage and use the resource in a sustainable way if they are free to work out the arrangements for themselves by trial and error. As Vernon Smith noted, "we have learned immeasurably" from such studies of "'natural' ecological experiments" (2003, p. 470, n. 21).

Simon's (1996) analyses of long – several centuries in some cases – time series of diverse resource prices that provide many natural experiments on how people respond to shortages – price increases – provides more general evidence on the subject of so-called limited resources. People develop more efficient technologies and more and better alternatives to the extent that – in tandem with increasing population – the real prices of all resources have trended downward over the long term.

Scientists can and do argue about what a particular experiment proves or disproves, if anything, due to uncertainty over the effects of possible confounding influences. Such disagreements are best addressed, however, by more and better experiments, as Smith (2002) argued.

9.4   Collecting Data

Scientists should try to ensure that their data are valid and reliable. To the extent that they are not, they should describe the problems with the data. Furthermore, they should use all data that can be shown to be valid and reliable, and *nothing more*. We stress "nothing more," because machine learning methods are susceptible to deriving models that include "predictor" variables for which there is no evidence of a causal relationship with the variable of interest.

9.4.1   Obtain All Valid Data

> I never guess. It is a capital mistake to theorize before one has data.
> Insensibly one begins to twist facts to suit theories, instead of theories
> to suit facts.
>   Sir Arthur Conan Doyle

By valid data we mean data that are from properly designed experiments or from observations of phenomena made in ways that are objective and consistent with domain expertise.

Explain the search procedures used to select data and explain the strengths and weaknesses of the data selected. Include all relevant data in your analysis as long as no substantive changes were made in the definition of the data or in the measuring instruments that cannot be properly adjusted for.

There is often more than one set of potentially useful data, and seldom one that is unbiased in any way. When there is more than one set of *valid data*, analyze each separately, then combine across the analyses to help control for biases. For example, data on country A's exports of product X to country B may differ from country B's data on imports of product X from country A. In such cases, averages of the two series can help to reduce biases.

When searching for evidence on causality, Freedman (1991) demonstrates the value of prior thinking about causes followed by direct observation. For example, John Snow's discovery of the cause of cholera in London in the 1850s came about from "clarity of the prior reasoning, the bringing together of many different lines of evidence, and the amount of shoe leather Snow was willing to use to get the data" (induction) in his survey of the location of the deaths and their relationship to neighborhood water pumps.

Surveys of people's intentions or expectations can be useful for situations in which people normally make plans that they can fulfill or have the experience necessary to assess how likely they are to do the thing of interest. The difficult part is how to ask questions so that people understand them. Even small differences can lead to substantial changes in the answer. For example, asking, "how tall was the basketball player?" produces different answers than asking, "how short was the basketball player?" Researchers often underestimate the number of revisions needed to produce a questionnaire that can provide valid data. Excellent advice on question wording and on procedures for asking questions is available in Dillman et al. (2014).

9.4.2 Ensure That the Data Are Reliable

Once the data are found to be valid, one should check for reliability. Reliability means the degree to which the results agree when the same collection procedures are repeated. For example, if the measure is based on expert judgments, are the judgments similar across judges? Alternatively, are the same judgments made when the same expert repeats the judgments over time? Have the measuring instruments changed over time? Are the measurement instruments in good working order? Have any unexplained revisions been made in the data?

Reliability can be improved by using larger sample sizes. However, assuming the samples are representative, there are diminishing returns as the sample sizes increase. For example, a study of 56 political polls – with samples varying from 575 to 2,086 – found that the accuracy of the poll data had little relationship with sample size (Lau, 1994).

9.5 Analyzing Data

Scientists are responsible for using appropriate methods for analyzing data. Describe your methods and models before starting the analysis, and record subsequent changes, if any, to the data or procedures.

If your findings are unexpected, consider that they might be the result of an error *and* that you might have stumbled on something important. Check your analysis to protect against the former and, if you find no error, design further studies to test hypotheses about the unexpected finding.

9.5.1 Use Models That Incorporate Cumulative Knowledge

The form of models used in analysis should reflect cumulative knowledge about relationships. For example, incorporate all variables known to be important, and use logarithms of the variable values if the relationship is known to be of the form that a percentage change in one is expected to result in a percentage change in the other.

On the other hand, if cumulative knowledge on relationships is such that the directions, sizes, or natures of relationships are uncertain, avoid specifying models that rely on assumptions about relationships that would bias the analysis.

9.5.2 Use Simple Methods

As we noted in Chapters 2 and 6, academics, bureaucrats, and clients often prefer complex methods. If you are devoted to making useful scientific discoveries, however, you will find that Occam's Razor is an important principle.

9.5.3 Use Multiple Validated Methods

Scientists are responsible for providing evidence that their methods have been validated for the purpose for which they have been used. When more than one method is valid, using multiple methods is likely to increase the validity of the findings.

Common usage of a method is not evidence that it is valid. For example, the statistical fit of a model to a set of data – such as might be measured by the adjusted-R^2 statistic – is not a valid way to assess the predictive validity of the model. Numerous studies have reached that conclusion. Six such studies were described in Armstrong (2001a, pp. 457–458).

9.5.4 Draw Conclusions Only on Practical Importance of Effects

Knowledge of effect sizes is vital for developing rational policies. To estimate effect sizes for causal relationships, researchers should use domain knowledge and prior experimental studies. Then combine the estimates from domain knowledge and any prior experimental studies with estimates from your own research for each effect size.

Does the effect have a practical importance? Is the effect from treatment "A" an improvement on the effect from treatment "B"? Would policy "X" increase welfare over the long term compared to laissez faire? The logical way to determine practical importance is to undertake a comprehensive cost–benefit analysis based on effect size estimates ignoring, as always, the invalid criterion of statistical significance.

10 HOW SCIENTISTS CAN DISSEMINATE USEFUL FINDINGS

In this chapter, we are concerned with how to inform those who could use your findings. There is little point in doing useful scientific research if potential users of the findings are not aware of them, or if they cannot understand the findings or how to use them.

10.1 Writing a Scientific Paper

Start writing the working paper early in the project, and keep copies of drafts as a record. Doing so will help in providing full disclosure. It also helps the writer to start writing early, as the subconscious helps in solving problems.

10.1.1 What to Include in a Scientific Paper

Checklist 10.1 (Content of a scientific paper), overleaf, provides a summary of the following guidelines on writing a scientific paper.

10.1.1.1 Explanation of Findings, and Why They Are Credible and Useful

The first thing to tell the reader is what your findings are, how you reached them, and why they are useful. Show how they can be used to, for example, improve decision-making, develop better policies, save lives, improve efficiency, reduce costs, reduce conflict among people, or identify important relationships. Describe the effect sizes of the findings,

Checklist 10.1 Content of a scientific paper

1. Explanation of findings, and why they are credible and useful	☐
2. Descriptions of prior hypotheses and any changes	☐
3. Descriptions of data and methods allowing assessments of validity, and replication	☐
4. Verified citations of scientific papers with substantive findings relevant to the evidence presented	☐
5. List of those who provided author-solicited peer review	☐
6. List of funders who expect to be acknowledged	☐
7. An oath on responsibility and disclosure	☐

and what the magnitude of the effect means for practical purposes, in plain English.

Authors must convince readers that their findings are useful. In other words, they should answer the "so what?" question. Do all of the above in a structured abstract. While that may seem obviously good practice, an examination of 69 papers in the 2000–2002 issues of the *International Journal of Forecasting* and 68 papers in the *Journal of Forecasting* found that only 13 percent mentioned the findings in the abstract. That occurred despite the journals' instructions to authors to do so (Armstrong and Pagell, 2003).

In the next chapter of this book (Chapter 11: How Stakeholders Can Help Science) we provide suggestions on standards for journals and recommend that editors enforce their standards by requiring compliance with a checklist.

Authors should assure readers that their findings are credible. One way to do this is to state that you used the criteria for compliance with the scientific method in your research and provide access to your completed copy of Checklist 3.1, *Compliance With Science Checklist*.

Don't try to make a silk purse out of a sow's ear: If you cannot show that your paper is useful, do not publish it. When Scott started his career, one of his first journal submissions involved sophisticated statistical analyses of a large data set. It was accepted by the leading journal in marketing. In the time between submission and acceptance, however, he became skeptical that the findings, while technically correct, were of any use, so he withdrew his name from the paper.

Scientists should write clearly so that all those who might be able to use the findings of their research can understand what to do. Use common words to describe everything. Avoid jargon. If you must use jargon, explain it on the first use. Crichton's (1975) list of 10 recurring faults in academic writing – see Chapter 6 – can be used as a checklist of what to *avoid* in writing a paper in order to communicate useful findings.

Mathematics is only a language; it is not part of the scientific method. Mathematical proofs do not provide scientific proof. Use mathematics only when needed to clarify your explanation and to explain the magnitude of effects. Avoid complex mathematical notation. If the math is complex or lengthy and you believe it will help some readers, put it in an appendix available on the Internet.

Round the numbers to make the paper easier to read and remember. Doing so also avoids implying a degree of precision when that is not justified.

See Tufte (2001) for advice on the graphical presentation of data. For example, he recommends against the use of pie charts (Tufte, 2001, p. 178).

When editing a paper, use hard copy. A meta-analysis of 76 comparisons of reading from paper versus reading from a screen from 54 studies involving more than 170,000 participants provided effect size estimates. The authors concluded that comprehension was greater when the reader is reading from paper (Delgado et al., 2018). Another meta-analysis involving 33 experimental studies from 29 reports found that comprehension, retention, and reading speed were better with hard copy than on screen (Clinton, 2019).

Revise often to reduce errors, improve clarity, and reduce length without eliminating important content. Typically, we revise our papers more than 50 times: the more important and complex the problem, the more revisions.

10.1.1.2 Descriptions of Prior Hypotheses and any Changes

Describe the various hypotheses and the conditions under which they apply. Also describe the criteria that will be used to compare the hypotheses.

Describe how the various hypotheses compare with one another, based on evidence prior to the current study.

During the course of the research, new evidence might affect the hypotheses. Describe in the paper any revisions to the hypotheses.

10.1.1.3 Descriptions of Data and Methods Allowing Assessments of Validity, and Replication

Describe how you searched for cumulative knowledge, selected from available data, designed experiments to obtain new data, and analyzed data using validated methods. Address issues that might cause concern, such as ensuring that any subjects would not be harmed.

Researchers are responsible for deciding what to report. They best understand what information should be included in a paper for publication, and what should not. For example, do not include information that would be useless, harmful, confusing, or misleading.

Researchers should keep a log or save all working versions of the paper to track important changes in hypotheses or procedures. For example, you might find a newly published paper that provides evidence that suggests you should add a causal variable to your study. That could be recorded in a log. Use of a log can also resolve disputes as to who made a discovery. Alexander Graham Bell's log of his experiments on telephony had a two-week gap, after which he described a new approach that was almost identical to an application to the US patent office by another inventor on the same day. After Bell was awarded the patent, the other inventor sued, but the courts concluded there was not sufficient evidence to convict Bell of stealing the idea described in the patent. Many years later, the missing pages from Bell's log were found leading to the conclusion that Bell had stolen the invention (Shulman, 2008).

While disclosure of *relevant* information is essential for science, regulation is not the answer. Reviews of experimental studies found no evidence that government mandated statements helped the people they were intended to help. On the contrary, they confused people and led them to make inferior decisions (Ben-Shahar and Schneider, 2014; Green and Armstrong, 2012).

Science does not need mandatory disclosures and disclaimers. Those who are skeptical of a study's findings can undertake a direct replication. If the authors of the original study fail to provide the necessary information when asked, the skeptical researcher can report that as a failure to comply with the scientific method. Nothing more need be said.

Moreover, government mandated messages violate scientists' freedom of speech. Consider a scientist who needs funding to run experiments to evaluate a controversial government policy. Funders might be willing to help but not if doing so would subject them to public censure, protests, boycotts, or death threats. Thus, scientists should be free to state that they received funding for their research from an anonymous donor.

Given that violations of the scientific method are common, authors might also want to assure readers by including an "oath" in their papers, affirming that they have complied with the scientific method. We suggest that this be done with a statement that the authors "complied with the *Compliance With Science Checklist.*" In addition, mention all those who provided peer review prior to submission who are willing to be named.

10.1.1.4 Verified Citations of Scientific Papers with Substantive Findings Relevant to the Evidence Presented

Do not cite advocacy studies as evidence. Doing so violates the scientific method.

Never cite findings from a study unless you or a co-author has read and understood the document in which it was described. Doing so would be unethical. We suggest including a statement in your paper verifying that at least one author has read each of the cited works.

When claiming that substantive findings support a point you are making, describe the evidence contained in the cited paper. That will enable the reader to judge what to expect from the paper. Avoid mysterious citations – that is, citations that provide no description of the findings in the cited paper. If a cited paper provides only opinions, make that clear to readers.

To check that you have properly summarized substantive findings from a paper, send the relevant text from your paper to the authors of the paper and ask if you have described their findings correctly. Also, ask if you have overlooked any papers with relevant scientific findings, especially if they conflict with your conclusions. If a response is not forthcoming, send a polite reminder.

By following that practice, we have found that many researchers reply with important corrections or suggestions for improvements, along with references for relevant studies that we have missed. As a result, our mistakes have been reduced, descriptions improved, and relevant papers identified. Inform readers that you have conducted a survey of cited authors and mention the number and percentage of contacted authors who replied.

Provide electronic links (URLs) to the papers you cite so that readers of an electronic copy can easily check the strength of the evidence they provide. Including links in working drafts of a paper also helps authors and co-authors to check their summaries of findings as they refine their own descriptions.

Researchers own the copyright of their last working paper prior to publication, and can create a PDF file of it for posting on the Internet and for distribution by email. Given that in recent times journals have seldom contributed substantially to revising a paper – it used to be common in the fields that we contribute to that journals would be responsible for editing and proofreading – the content of the working paper will likely be very close to the version published by the journal.

10.1.1.5 List of Those Who Provided Author-Solicited Peer Review

Obtain many peer reviews prior to submitting a paper for publication, especially if you believe the paper has important scientific findings. We suggest ten or more peer reviews to ensure that almost all errors have been found. Frey (2003), for example, received help from 50 reviewers.

Grade yourself on how many suggestions you were able to use from each reviewer. That applies also to journal reviews, even if your paper is rejected: try to extract ideas for improvements to your paper from the comments and suggestions in the reviews.

In our experience, scientists tend to respond to requests for help if they believe the problem is important and if they can contribute. Also, they will want to know about your paper if it makes an important contribution in their area of interest, and if it might lead to fruitful collaborations as it did for the authors of this book nearly 20 years ago.

Send a personal message to each scientist you identify as having relevant interests and expertise asking for suggestions on ways to improve the paper. If a paper is important and relevant to them, scientists are likely to be willing to help. For example, when Scott was an unknown assistant professor, he sent a paper to Nobel Laureate Milton Friedman asking if he would provide a review. The paper provided experimental evidence related to Friedman's interest in corporate social responsibility. Friedman promptly responded with a helpful review. In our experience, mass appeals for reviews, such as posting a working paper on the Internet or sending it to mailing lists, lead to few useful suggestions.

The process of preparing and delivering the paper to an audience often proves useful by encouraging one to anticipate objections. On the other hand, it is difficult to get useful suggestions during live presentations of findings. We suggest distributing a page for comments at the start of your talk, asking for comments, and ending a few minutes early so people in the audience can write suggestions for you.

Ask for ways to improve your paper. People respond differently in a helping role than they do when they are asked to *evaluate* a paper. Provide a link to a copy of your talk on the Internet to encourage follow-up suggestions.

Use editors to improve the clarity of the writing and reduce length. We typically use at least two copy editors for each paper, and more for books.

Acknowledge all those who provided reviews, *and* those who made useful suggestions.

10.1.1.6 List of Funders Who Expect to be Acknowledged

If a funder prefers to remain anonymous, state that in the paper. In such cases, it is imperative that sufficient disclosure is provided that any researcher who is concerned that funding might have influenced the study's findings can replicate the research.

10.1.1.7 An Oath on Responsibility and Disclosure

For example, "The authors made all decisions about the study design, data, methods, and writing. Methods and data are fully disclosed in order to allow checking and replication."

The 25 items in Checklist 10.2 (Writing a scientific paper) are based on principles from a review of the literature on persuasion (Armstrong, 2010).

10.1.2 Avoid Distractions When Writing

Write alone in a quiet area. If you need to play music, avoid music with lyrics. One experiment – involving 56 male and 46 female subjects – found that music with lyrics reduced attention to the task at hand by 7 percent relative to no music and by 2 percent relative to music with no lyrics (Shih et al., 2012).

10.2 Disseminating Findings

Benjamin Franklin advised researchers with important scientific findings to take on the responsibility of disseminating them. A promising place to start is with a good university library. For example, the University of Pennsylvania library initiated "Scholarly Commons" to make it easier for people to access the papers of the university's researchers. The library also provides regular reports on readership. For example, Scott's cumulative number of downloads on Scholarly Commons alone as of September 2019 was over 560,000. That figure does not include the number of people who read the papers in the journals in which they were published, nor of readers of papers downloaded from one of the many other repositories.

Other universities have taken similar approaches. For example, the University of South Australia's library maintains a Research Outputs Repository that provides open access versions of the university's researchers' work.

Other routes for dissemination – including journalists, authors of pop-science books, and textbooks – can also be effective if they describe the evidence.

10.2.1 Publishing

The traditional route to dissemination is via scientific journal articles. Publishing an article in a journal provides a kind of certification for your findings. As we have shown above, however, traditional journals and their reviewers suffer from confirmation bias. Thus, high status journals can be reluctant to publish important new findings that challenge the conclusions of previously published works.

If you believe your paper has useful scientific findings, you might consider sending a proposal or a partly finished paper to editors of relevant journals and ask if they would invite your paper. By following that approach, you have not formally "submitted" your paper; thus, you could make the offer to a number of journals at the same time – you should inform the editors that you are doing so,

Checklist 10.2 Writing a scientific paper

1. Make the first word in the title descriptive, avoiding adjectives, including unnecessary articles (like "the" or "a/an")	☐
2. Use a short explanatory title describing your findings	☐
3. Use past tense to report findings to avoid implying that the issue is settled	☐
4. Provide a structured abstract (see Checklist 11.1)	☐
5. Use an introduction to let readers know what to expect	☐
6. Use descriptive headings to guide readers	☐
7. Use numbers or letters for three or more items in a list	☐
8. Use examples to *illustrate* findings, *not as evidence*	☐
9. Organize tables and charts so conclusions are obvious	☐
10. Avoid pie charts	☐
11. Avoid colors, unless informative and necessary	☐
12. Be specific, using words with concrete meaning	☐
13. Avoid negative words for ease of understanding	☐
14. Use short sentences and avoid unnecessary words	☐
15. Avoid uncommon words, unless explained	☐
16. Break text into paragraphs that contain one idea each	☐
17. Describe how each substantively cited work provides evidence (i.e., avoid mysterious citations)	☐
18. Cite a source for evidence only if it has been read by at least one of the authors	☐
19. Put citations near at the end of sentences	☐
20. Use a common *serif* typeface with black-on-white text	☐
21. Use a calm tone, avoiding exclamation marks and uppercase in the text	☐
22. Use footnotes sparingly	☐
23. Rewrite the report until it is clear and interesting	☐
24. Use editors to improve clarity	☐
25. Proofread to eliminate errors	☐
TOTAL	[ ]

however. If an editor agrees to your proposal, you could also offer to obtain reviews.

The lead times for adoption of useful findings are often long, even when they have been published in high-status journals. For example, Meehl's (1954) finding that quantitative models are superior to expert judgment for personnel selection was widely cited and replicated by many other studies, but almost half a century elapsed before it gained acceptance by a renegade baseball manager. He was widely scorned at the time, but most professional sports teams have now adopted the approach, where it has provided much better results at a lower cost than expert opinions, such as those of baseball scouts (Armstrong, 2012b; Lewis, 2003).

Unfortunately, outside of sports teams, few organizations use Meehl's findings. For example, they seem to be ignored by consulting firms that charge large fees to do executive searches on behalf of corporations in the United States (Jacquart and Armstrong, 2013). In addition, we are aware of only one university that has used Meehl's approach. Scott's attempt to introduce the approach at a university failed.

The eight items that follow provide practical guidance for publishing useful scientific findings. They are summarized as Checklist 10.3, below. They are based on common sense and domain knowledge.

The publication of useful scientific knowledge can be achieved without journal publication by posting papers on the Internet at no cost on websites such as ResearchGate. For example, one of our papers on ResearchGate – "Forecasting Methods and Principles: Evidence-Based

Checklist 10.3 Disseminating useful scientific findings

1. Provide thorough responses to journal reviewers	☐
2. If a paper with useful scientific findings is rejected, appeal to the editor	☐
3. Publish in *PLoS ONE* or similar if you meet their criteria	☐
4. Publish a working paper on ResearchGate or similar	☐
5. Publish research findings in a book	☐
6. Directly inform those who can use your findings	☐
7. Cooperate with those who want to replicate your study	☐
8. Publish corrections for mistakes found after publication	☐

Checklists" (Armstrong and Green, 2018) had been "read" more than 30,000 times by December 2021, and it continues to be read on ResearchGate after being published in a journal.

10.2.1.1 Provide Thorough Responses to Journal Reviewers

Authors should take responsibility for making all changes that would improve their paper. They should, however, avoid making changes that they consider would harm their paper.

10.2.1.2 If Useful Scientific Findings Are Rejected, Appeal to the Editor

If your paper has useful scientific findings and it was rejected by reviewers, appeal to the editor of the journal. Provide detailed point-by-point responses to journal reviews.

A study of the *American Sociological Review* found that only 13 percent of the authors who complained had their papers accepted by the journal (Simon et al., 1986). However, objecting has worked well for Scott. Nearly all of his most useful papers have been rejected on first submission with no encouragement to resubmit. After revising the paper based on the reviews, he would provide reasons to the editor why the paper should be published. For some papers, that involved doing further experiments. In general, the revisions were upsetting to the reviewers. However, Scott has persisted with that approach and eventually all but two of his most important papers were published in reputable scientific journals thanks to the editors. On the other hand, the mandatory review process has wasted an enormous amount of our time, as it no doubt has for other authors.

10.2.1.3 Publish in *PLoS ONE* or Similar if You Meet Their Criteria

The journal ranking systems used by universities create long lead times for publishing in "top" journals, and low probabilities for acceptance. Consider alternatives, such as *PLoS ONE*.

PLoS ONE will publish your paper for a fee if you comply with their criteria. It is a well-accepted way to publish your paper, and has proven to be a good way to disseminate useful scientific knowledge to a large number of people.

10.2.1.4 Publish a Working Paper on ResearchGate or Similar

Thanks to the Internet, scientists can self-publish a working paper on ResearchGate for free and without censorship. That establishes a claim to your findings, and it can lead to high readership, as it is, in effect, free open access to everyone. According to Wikipedia, ResearchGate "is the largest academic social network in terms of active users."

ResearchGate publishes working papers and other works when the authors submit them. Authors can make changes in the papers when they like, such as to correct errors or to add evidence.

The final version of a paper can still be published in a traditional journal. Authors own the copyright to the version of the paper that was submitted to the journal that published the paper. In most cases, the papers are very similar to the published version. One study, however, provides a caution to authors, finding that of a sample of 500 papers on ResearchGate, 40 percent appear to infringe copyright because the authors uploaded a final published version (Jamali, 2017).

Unlike journals, ResearchGate makes papers freely available to potential readers outside of academia. Authors can also make revisions at later dates. For example, an early version of this book was posted as a working paper on the Internet in July 2016. By December 2021, it had accumulated nearly 14,000 "reads" – views of the publication on the Internet.

Scholarly Commons by the library of the University of Pennsylvania – and other schools – gets enormous readership by academics. That has been the source of many downloads of our papers.

In another case, we were invited to write a chapter for a book. After many revisions, the editors said that our chapter was not a good fit with their book as it was "too technical," so we posted it as a working paper version on the Internet in 2012 (Armstrong and Green, 2012). We planned to revise and submit it as a paper elsewhere, but never did so. Despite not having been "published," it had been cited 128 times by December 2021 according to Google Scholar.

Open access is a wonderful innovation. But it has also led to the creation of fake or "predatory" journals. They have names that appear scientific, offer fast but useless reviews, and open access, but with high fees. Beware of publishing in or citing articles in fake journals. A useful web site is *Beall's List of Predatory Journals* at https://beallslist.net/ – based on Beall (2012) – which provided a list of over 1,100 such journals as of January 2018. For an account of the activities of predatory journals, see "Paging Dr. Fraud" in the *New Yorker* (Burdick, 2017).

10.2.1.5 Publish Research Findings in a Book

Unlike journal editors, book publishers have little interest in protecting current beliefs. They are motivated to make money by publishing interesting and useful books. As a consequence, books, book chapters, and monographs can provide ways to avoid censorship and to publish useful findings that might be unpopular among the editors of journals.

Books offer an opportunity to provide a complete review of scientific research on a given problem, along with full disclosure and, in general, with little need to satisfy publisher appointed reviewers. In addition, books allow authors to disseminate useful new scientific findings. Nevertheless, scientific books are time-consuming for authors to write, so they are not an attractive option for young researchers in need of publications to gain tenure.

Compared to journals, the market for books seems to be more open. Also, instead of having to submit a paper to one journal at a time, authors are able to make simultaneous submissions of proposals or drafts to book publishers. That increases the likelihood that authors can negotiate terms in the contract.

We advise scientists against writing pop-science books. In our judgment, most of them fail to summarize prior findings properly. In addition, they often fail to inform readers about the original sources, and casually mix apparent facts with opinion and speculation. Finally, their sources often rely on unscientific papers. Some books have, nevertheless, overcome those challenges and been successful at disseminating useful scientific knowledge. One example is Cialdini's *Influence* (2001). Another is Gigerenzer's *Simple Heuristics That Make Us Smart* (Gigerenzer and Todd, 2000). They succeed largely, in our opinion, because they reported findings from their own scientific research and on related research by others.

10.2.1.6 Directly Inform Those Who Can Use Your Findings

If your findings are useful, you should try to gain the attention of those who might be able to use them. Use titles that are descriptive and understandable. Lead with the most descriptive words. Avoid adjectives, including articles (i.e., "the" or "a/an") as the first word in the title; advice that we regret having violated in the past. Avoid obscure titles, such as the following title of a paper published in a scientific journal in 2018: "Quantile estimators with orthogonal pinball loss function."

Historically, some of the dissemination of findings from scientific papers was expected to occur via classrooms in universities. That appears to have worked in some disciplines, such as medicine and engineering. However, it currently appears to be rare in many university departments that are typically thought of as being concerned with science.

Make your paper easy to obtain. Consider journals that support Green Open Access policies, whereby the paper is put online for free after a certain period. Alternatively, authors can pay for Gold Open Access whereby the paper can be freely downloaded as soon as it is published. Open Access is rapidly expanding. See, for example, the Harvard Open Access Project at http://cyber.law.harvard.edu/hoap for the latest developments.

Do not despair when your most useful papers are cited less often than your other papers. A survey of 123 of the 400 biomedical scientists whose papers were most cited during the period 1996–2011 revealed that 16 percent of them considered that the paper they regarded as their most important was not among their top 10 for citations. The leading reason appeared to be a bias toward papers that confirm current beliefs: 14 of those 20 scientists considered their most important papers to be more disruptively innovative or surprising than their top 10 cited papers (Ioannidis, 2014).

Face-to-face discussions are one of the *least* effective ways of changing opinions. In such situations, those who take part tend to invest their energy in defending their opinions and thus come away with a strengthened belief in their current opinions. Studies on the topic are summarized in Armstrong (2006).

The number of times a paper *with useful findings* has been read and cited provides a good measure of dissemination. The primary responsibility for disseminating useful research findings falls upon the researchers. Researchers have the copyright to the working paper that they submit to a journal, the right to post it on their own website and on repositories such as Google Scholar, Kudos, ORCID, RePEc, ResearchGate, Scopus, SSRN, Web of Knowledge (previously ISI), and others. Researchers should also send copies to colleagues, researchers cited in important ways, and to others who helped on the paper.

10.2.1.7 Cooperate with Those Who Want to Replicate Your Study

Replication is vital to science, *and useful to the scientists whose work is being replicated*. Why then, have many researchers reported failure to cooperate? We are not sure, but we speculate that alerting the publishing journal's editor-in-chief to the problem and requesting that the journal's disclosure policies be enforced may help. If cooperation is still not forthcoming, the journal should alert readers if an attempt has been made to replicate the paper and that the authors have not cooperated, along with their reasons for not cooperating, or withdraw the paper.

10.2.1.8 Publish Corrections for Any Mistakes Found After Publication

If others tell you confidentially about possible mistakes, thank them profusely! Why? When the authors make corrections in a published paper, readers appreciate that and view the author in a favorable manner. However, when journals make even a single retraction that is not initiated by the author, the credibility of the author's body of work often suffers, resulting in a decrease in citations (Lu et al., 2013). If corrections are needed, ask the journal to make the corrections. If the mistakes are serious, ask the journal to retract your paper and explain why.

If you find what seem to be mistakes by others, send them a *personal message*. Everyone gains. Should the authors fail to respond, and the findings are important, you might want to report your concern to the journal in which the paper appeared, or to replicate their study.

If you are concerned that a paper you have cited has been retracted, you might find a discussion on the paper on the "Retraction Watch" website, available at https://retractionwatch.com/.

Our concern here is with whether papers include seriously flawed scientific findings. We fail to understand how authors can harm

Checklist 10.4 Preparing a talk on scientific findings

Checklist items are based on evidence from Armstrong (2010) or are logical or based on expert judgments.

Organization

1. Use a single theme to organize your scientific findings	☐
2. Describe objectives of the talk, including action-steps	☐
3. Build the presentation around your scientific findings	☐
4. Show evidence for your findings	☐
5. Use two-sided arguments describing risks, limitations	☐
6. Avoid jargon and uncommon words	☐
7. Plan to take less time than is available to allow for interruptions and problems	☐
8. Have additional slides in reserve in case there are a few questions or you have more time than expected	☐
9. Rehearse: If the talk is important, present your talk to one or more people acting as your intended audience	☐
10. Prepare a hard copy agenda for your audience	☐
11. Provide a link to your paper or slides	☐
12. Prepare hard copies of your slides in case of problems	☐

Visuals

13. Use simple visual aids, especially for data	☐
14. Synchronise oral and visual parts of the talk by using animations that introduce one point at a time	☐
15. Keep to 10 lines or fewer of text for most slides	☐
16. Eliminate anything from visuals that contains no useful information (e.g., wallpaper or color)	☐
17. Use high contrast (e.g., black text on white background) to enhance legibility	☐
18. Use a *sans serif* font to enhance legibility	☐
19. Provide informative titles for each exhibit	☐
TOTAL	[ ]

science by "plagiarizing" their own works. For example, when Scott was asked many years ago if he would allow one of his papers to be published in a new journal, he asked both journals editors and both said it was fine by them. In addition, in the latter half of the twentieth century, it was popular to reprint journal articles in books of "collected readings" on a topic so as to get additional readership and permission was freely granted. During those years, Scott had 30 papers reprinted, some more than once, for a total of 60 re-prints. To the extent that the papers have useful findings, this seems beneficial for science, and certainly it was not unethical. The notion that one cannot publish the same thing in different places should be based on the agreements among authors and publishers.

Some years ago, Scott learned that his work had been plagiarized. He called the author – who was a young PhD – and told him that

Checklist 10.5 Making an oral presentation

1. Use one speaker	☐
2. Ask the audience to write suggestions for improvements	☐
3. Answer only clarification questions during the talk	☐
4. Acknowledge non-clarification questions, and undertake to address them at the end of the talk	☐
5. Pause for two seconds before key points to create interest	☐
6. Pause after key points to allow people time to reflect	☐
7. Pose questions, pause, then answer your own question in order to gain attention	☐
8. Make eye contact to raise interest and increase trust	☐
9. Avoid humor so as not to distract from the talk's content	☐
10. Repeat key points by rephrasing them	☐
11. Orient questions toward improving the paper	☐
12. Ask for clarification if uncertain about a question and offer to discuss after the talk if you don't have an answer	☐
13. Avoid answering questions that need your further consideration; note the questions and respond later	☐
TOTAL	[ ]

he was not going to do anything because there was no harm to science, but suggested that his career would go better if he did not plagiarize in the future as other authors might not be willing to let it pass.

Never publicly accuse a scientist of unethical behavior. Even if you are correct, it may cause harm to you. If you happen to be wrong, it will be worse. Be kind; contact the author and suggest why you think there might be an important error.

10.2.2   Preparing a Talk and Giving a Presentation

Checklist 10.4 is for preparing a talk on scientific findings. It includes some suggestions that may seem obvious. Completing the checklist will, however, help you to follow the guidelines consistently.

Checklist 10.5 provides guidance for oral presentations.

11 HOW STAKEHOLDERS CAN HELP SCIENCE

This chapter addresses how stakeholders of scientific research can assess whether research is compliant with the scientific method and then promote useful scientific research to improve products, services, processes, methods, and decision-making. We address stakeholders in sections for universities, scientific journals, governments, regulators and courts, and media and interested individuals.

11.1 Universities

> In the customs and institutions of schools, academies, colleges, and similar bodies destined for the abode of learned men and the cultivation of learning, everything is found adverse to the progress of science.
>
> Francis Bacon (*Novum Organum*, 1620 [1863])

Important contributions to science depend heavily on skeptical scientists. Yet skepticism is under threat in many universities. Students and academics in the United States, Australia, and the UK actively try to prevent scientists who have conclusions that challenge their own beliefs from speaking on campus by calling for the university to cancel their lectures or by disrupting their talks.

11.1.1 Hiring Scientists

Few have the general mental ability (GMA) to become scientists and, of those who do, few have the right personality to be a scientist.

Thus, it seems cost-effective for organizations to try to identify those most capable and motivated to succeed in a career as a scientist.

The small fraction of people who have the ability and desire to do useful scientific research amounts to a number that is almost certainly smaller than the number of people currently employed as academics in universities around the world. As a consequence, universities cannot expect to compete with each other and with industry and other research institutions to populate their academic ranks with researchers doing useful research. Rather than try to turn all academics into researchers, better to hire those who do useful research as researchers, and to take that burden away from other academics.

As we discussed in Chapter 8 (What It Takes to Be a Good Scientist), a researcher who has a good PhD is likely to have a sufficiently high GMA to have the *potential* to succeed as a scientist. GMA is the single best predictor of success in any job. If a PhD is not a requirement for the job, then screen out candidates with an IQ of less than 130.

Schmidt et al.'s (2016) meta-analysis of 100 years of research provides a summary of evidence on the predictive validity of 30 personnel selection methods, including GMA. For candidates who meet the GMA requirement, integrity tests and structured interviews each markedly help to improve predictions of job success. Other measures do not, either because they lack individual validity (e.g. person-organization fit, age, and agreeableness), or because they correlate strongly with GMA and so have little incremental validity in predicting success (e.g. job knowledge tests, peer ratings, and assessment centers).

Integrity tests – which are used to *avoid* hiring employees who are more likely to indulge in counterproductive behaviors on the job – seem to provide the single largest incremental increase in job success predictive validity. The tests also measure the qualities of conscientiousness – we provided a test for self-control in the form of Checklist 8.1 in Chapter 8 – emotional stability, and honesty–humility.

Applicants should be told what criteria will be used to evaluate their applications for a position as a research scientist. For hiring researchers with a track record – at least a PhD thesis if not a published paper – we suggest using the *Compliance With Science Checklist* (Checklist 3.1) to assess their contributions to useful scientific findings.

In general, stick with objective announced criteria and avoid unaided judgments about a candidate's suitability. Meehl's (1954) book, *Clinical Versus Statistical Prediction*, described four decades of research on the value of judgments versus statistical models; the models provided more accurate predictions than did the unaided judgments. An updated meta-analysis by Grove et al. (2000) reached the same conclusion.

11.1.1.1 Evidence of Prior Useful Research

Relevant experience is important for researchers. So why not ask applicants for positions as researchers to describe what useful scientific discoveries, or entrepreneurial ventures, or creative endeavors they have been involved in.

As a visiting professor at the University of Canterbury in Christchurch, New Zealand, in 1985, the Research Provost told Scott that when he asked researchers to describe what useful findings they had discovered over the past year, they went on at length to describe what they were *doing* in their research, but rarely discussed what they *had discovered*. Universities could be more effective at recruiting scientists who make discoveries if they assessed candidates' records using that criterion.

Ask candidates to submit their most useful scientific paper to an administrative assistant who would redact the author's name, education, age, and other demographic information.

In the first test of the inter-rater reliability of Checklist 3.1, *Compliance with Science*, we independently rated the best papers submitted by applicants for positions as assistant professors in marketing science. Most of the papers complied with at least one of the scientific criteria.

Avoid hiring quotas irrelevant to scientific research. Scientific research requires unusual skills. In that sense, it resembles sports. If you do not have the aptitude, you will not be successful, no matter how hard you practice. Would a university select football players on the basis of being representative of the general population? A meta-analysis found that bio-demographic diversity does not improve the performance of groups, either (Horwitz and Horwitz, 2007).

We suspect that the Agricultural and Industrial Revolutions would have been less bountiful and productive had researchers been chosen based on quotas to meet demographic diversity targets.

11.1.1.2 Contracts with Scientists

Require that scientists engaged by the organization comply with the scientific method in all their research. The contract should also include a clause stating that compliance will be monitored.

11.1.2 Freedom to Think and Speak

In 1911, a Wharton School economics professor, Scott Nearing was fired for advocating socialism. That led to protests in the United States. As a result, some universities adopted "tenure" arrangements to protect academics from being censored by university managers by granting those who met their university's standards a permanent position.

By the late twentieth century, free speech was again under threat on campuses. In 1999, the Foundation for Individual Rights in Education (FIRE) was founded to defend the freedom of speech of students and academics at US universities. FIRE was founded in response to widespread support for the conclusion of a book written by a University of Pennsylvania professor and a civil liberties lawyer – that suppression of speech on campus is the norm. Their book was called *The Shadow University: The Betrayal of Liberty on America's Campuses.*

In 2020, a similar organization was established in the UK: The Free Speech Union, or FSU. The FSU undertakes to come to the defense of members who find themselves being attacked for exercising their legal right to free speech.

One way to deal with the pressures that university managers face to suppress unpopular speech and the challenging of strongly held views is for each university to develop a clear policy on free speech for academics, and to enforce that policy. Consider the University of Chicago's statement on free speech:

> Of course, the ideas of different members of the University community will often and quite naturally conflict. But it is not the proper role of the University to attempt to shield individuals from ideas and opinions they find unwelcome, disagreeable, or even deeply offensive. Although the University greatly values civility . . . concerns about civility and mutual respect can never be used as a justification for closing off discussion of ideas, however offensive or disagreeable those ideas may be to some members of our community. Zimmer et al. (2014, p. 1)

As of September 2019, more than 60 colleges and universities in the United States had signed a statement adopting the University of Chicago free-speech protections. On March 21, 2019, President Trump issued an executive order that requires universities to allow "free inquiry" in order to receive federal grants.

The suppression of speech on campuses is not a uniquely United States, or UK, phenomenon. Take the recent example of marine scientist Peter Ridd at James Cook University in Australia. The university's management sacked him for criticizing colleagues for practicing advocacy research in relation to the health of the Great Barrier Reef. The case went to court, and the judge required little time to reach a decision. He simply noted that the university had a policy supporting free speech for academics (Ridd v. James Cook University, 2019).[1]

Universities can also help support researchers' freedom of thought and speech by refusing government funding and the regulation of researchers that goes with it. Funding from individual donors and private organizations such as corporations can provide funding to support the diverse interests and skills of researchers.

Consider Media Lab at MIT. Their research is supported by corporate sponsors and the funds are allocated among the researchers with no need for them to write grant proposals. While no program is perfect in every respect, the approach of asking for funding from private corporations seems reasonable, and diverse arrangements with different funders should make possible a diversity of research topics and approaches.

Another example of corporate support for scientific research is provided by the Ehrenberg-Bass Institute at the University of South Australia. The institute is named after two maverick researchers, and the institute's researchers and PhD students work on projects of their choosing with the help of generous financial sponsorship from many large corporations and access to data that is not commonly available

---

[1] At the time of writing, the university has successfully appealed the verdict. The judges ruled that the university's code of conduct constrains employees' employment agreement right to free academic expression. In early 2021 the High Court of Australia – Australia's highest court – granted Ridd leave to appeal. The appeal was heard and the court unanimously dismissed the appeal on October 13 on the grounds that his employment agreement did not provide for a general freedom of speech.

to other researchers. Scott and Kesten are both affiliated with the institute.

Has such a solution worked in the past? Yes. As mentioned earlier, wealthy gentlemen farmers supported the agricultural revolution, and British industrialists and their investors funded the research that created the Industrial Revolution.

11.1.3   Diversity of Ideas

If everyone is thinking alike, then no one is thinking.
Benjamin Franklin

Researchers, like most people, prefer to associate with those who have similar beliefs. That tendency has grown over the past half-century, such that political conservatives – in the United States sense of the term – are rare in social science departments at leading US universities (Duarte et al., 2015; Langbert et al., 2016).

An earlier survey of 266 members of the Society for Personality and Social Psychology found that 14 percent reported that they would "somewhat," or more than "somewhat," discriminate against politically "conservative" (US meaning) researchers in symposium invitations. Nineteen percent reported they would do so in paper reviews, 24 percent in grant reviews, and 38 percent in hiring decisions (Inbar and Lammers, 2012).

Universities could do much to foster skepticism and the diversity of ideas among academics by using explicit criteria, such as Checklist 3.1: Compliance with Science, in hiring decisions. In short, only criteria that are directly related to discovering and disseminating useful scientific findings should be used.

11.1.4   Foster Creativity

Creative people are motivated primarily by intrinsic rewards, including feelings of competence and autonomy. Extrinsic rewards can dampen the effectiveness of intrinsic awards for scientists. Much research has been devoted to the topic. See Gagne and Deci (2005) for a review.

Creative scientists are curious, willing to take risks, and persistent when faced with obstacles (Zhou, 2003).

Researchers benefit from an environment that stimulates creative problem solving. Evidence from natural experiments in medical research (179 studies) suggested that research that leads to innovation is mostly likely to occur in environments where researchers are free to design and conduct their research as they see fit *and* are assessed against the practical consequences of their innovations. The most innovative research was four times as likely to come from medical school and hospital researchers than it was to come from university researchers, and more than twice as likely to come from health agencies (Gordon and Marquis, 1966).

In a nearly 400-page survey of the history of innovation, Ridley (2020) concluded that freedom is essential for innovation to thrive. He wrote, "The main ingredient ... that leads to innovation is freedom. Freedom to exchange, experiment, imagine, invest and fail; freedom from expropriation or restriction by chiefs, priests and thieves; freedom on the part of consumers to reward the innovations they like and reject the ones that they do not" (p. 359).

Two field experiments involving 24 and 123 professional staff at a university and a for-profit hospital found that close supervision suppressed creativity (Zhou, 2003). The second experiment found that close supervision was particularly harmful to the creativity of staff who themselves had relatively uncreative personalities but who had a creative co-worker.

The Skunk Works division at Lockheed Corporation provided freedom for its scientists. The leader of the group requested and received a budget, and he allocated the money as he saw fit in order to do useful research. There was no interference by top management over how the money would be spent (Rich and Janos, 1996). The discoveries by that small group were revolutionary in the aviation industry, including the Nighthawk stealth fighter – which was "invisible" – and the Blackbird – the fastest plane ever made.

Workplaces can also make a difference. Because the presence of other people can reduce creativity, it makes sense to provide private spaces in the workplace.

In addition to fostering creativity, universities should, in the pursuit of objectivity, reject government grants. Instead, they should fund researchers who have a track record of useful findings and provide autonomy. For example, Scott has received much support from his departmental staff along with an annual budget to spend on his research. He supplemented his university budget with his own money, which in some years was a substantial contribution.

Private donations can provide additional funding. Donors may wish to specify what topic they would like to have researched, and they could insist on progress reports, which would help to encourage useful research.

11.1.5 Require Researchers to Report Useful Findings

Universities should ask their faculty to write an annual report that describes their useful scientific findings. The scientist can also use this as an opportunity to set objectives for their research. Edwin Locke and Gary Latham have conducted many experiments on the effects of objectives. Their findings provide strong evidence that setting objectives leads to substantial improvements for a wide variety of tasks when used in combination with good managerial judgment (Latham and Locke, 1979).

We believe that the use of Checklist 3.1 will make preparing annual reports more useful for researchers and for those evaluating their contributions to science. It would also eliminate the need for faculty to rely on criteria that are not related to the discovery of useful scientific knowledge, such as grant funding, number of papers published in highly rated journals, and citations of papers that do not include useful scientific findings.

11.1.6 Disseminate Useful Knowledge

Historically, universities have been responsible for disseminating useful scientific knowledge to students. That was typically accomplished through lectures and books.

To test dissemination of useful scientific findings in management, Scott, along with Randall Schultz, decided to assess the extent to which basic marketing texts – often with the word "principles" in the title – used scientific principles. Using nine basic marketing textbooks published between 1927 and 1989, four doctoral students found 566 principles, but none of the principles were supported by empirical evidence. Four raters agreed that only 20 of these 566 principles were meaningful. Twenty marketing professors rated the 20 meaningful principles as to whether they were correct, supported by empirical evidence, useful, and surprising. None met all the criteria and the professors judged 9 of the 20 principles to be nearly as correct when their wording was reversed (Armstrong and Schultz, 1993).

Interestingly, during that period, many papers with useful scientific knowledge were published, as Scott discovered when he examined research on persuasion. Early experiments were conducted by practitioners, and these were followed by academic researchers' experiments. Scott's book on persuasion rested on evidence from about 3,000 scientific sources. That prior research was the basis of almost one hundred evidence-based principles (Armstrong, 2010).

Universities could provide clearly written annual summaries of the useful scientific findings of their researchers. Researchers could lead the way by explaining – in simple language for all stakeholders – how their research has contributed to useful scientific knowledge. The findings could then be summarized by departments, schools, and universities.

Importantly, given that claims about causal relationships in published observational studies are often falsified by later experimental studies (see, e.g., Young and Karr, 2011) and that preclinical test results often fail to replicate in the lab or deliver the expected result in clinical trials (see, e.g., Begley and Ellis, 2012), universities should publicize the findings of attempted replications and extensions, including negative findings.

11.1.7 Employ an Ombudsman

Universities and other research institutions can create an ombudsman position. This enables people to seek help on a delicate situation while protecting their own identity. Thus, they could describe a suspected case of cheating, and the ombudsman would investigate. Scott, along with two colleagues, initiated such a position at the University of Pennsylvania after an unfortunate incident many years ago. The university took immediate action and the ombudsman position still exists.

11.2   Scientific Journals

Scientific journal editors and publishers could be doing more to encourage the production of useful scientific findings, and to disseminate those findings in a timely way to widest relevant readership. Our solutions in this section benefited from suggestions from Nosek and Bar-Anan (2012) and others. Some journals have already implemented some of the measures we describe.

11.2.1 Invite Papers with Useful Scientific Findings

In 2017, we examined a convenience sample of the aims and instructions to authors of six journals in the management sciences: *Management Science, Journal of Consumer Research, Marketing Science, Journal of Marketing Research, European Journal of Marketing*, and *Interfaces*. Only two attempted to explain that they were seeking papers with useful scientific findings. Those two did not, however, explain what that meant. Might it be that researchers seldom follow the scientific method because they are not asked to do so?

In their "Instructions to Authors," journals could ask for papers with useful scientific research, and provide incentives for such papers. At a minimum, the instruction could be operationalized as a requirement to test multiple reasonable hypotheses, where reasonable is defined broadly to include controversial hypotheses, non-trivial naive hypotheses, and the status quo hypothesis. While this is speculative, it has high face validity. That said, we are unaware of any journals where the instructions to authors explicitly or implicitly requested papers with useful scientific findings in the social sciences.

Might those who submit papers be more likely to submit papers with useful scientific findings if the journal stated that this would be the prime criterion used for deciding whether to accept a paper? In addition, might the resulting submissions include a greater percentage of papers with useful scientific findings?

The most direct way to assess whether a paper includes useful scientific findings is to ask the authors to explain the value of their findings in plain language. If they cannot explain why their findings are important, who can?

As we discussed in Chapter 4, the scientific method requires that papers with useful findings should be replicated and extended in order to assess the validity of the findings and to establish the conditions under which they apply. Journals can perform an essential role in the accumulation of scientific knowledge by inviting and publishing more replications and extensions of important papers. That practice could be encouraged by journal editors establishing a replications editor and a replications section, and inviting replications and extensions of important papers. Researchers on scientific practice have advocated such initiatives for decades (see, e.g., Evanschitzky et al., 2007; Hubbard et al., 1998).

Fidler and Wilcox (2018) optimistically concluded in their entry in the *Stanford Encyclopedia of Philosophy* on "reproducibility of scientific results," "[w]hile the meta-science has painted a bleak picture of reproducibility in some fields, it has also inspired a parallel movement to strengthen the foundations of science."

11.2.2 Require Structured Abstracts

We believe that the abstract is the most important part of a scientific paper. That is also the position of the American Psychological Association (APA) in their *Publication Manual* (2001, p. 12) "The abstract needs to be dense with information but also readable, well organized, brief and self-contained." If an abstract fails to make a clear case for the usefulness of the paper's findings, one would expect that the paper will be overlooked.

Structured abstracts have been standard practice for decades in medical science journals. Because structured abstracts can serve as checklists that can be easily monitored, it seems obvious that they would be substantially more useful than unstructured ones. Hartley (2003) provides evidence that they (a) contain more information, (b) are easier to read, (c) easier to recall, (d) easier for reviewers, and (e) welcomed by readers and authors.

The use of structured abstracts has led to a 30 percent increase in the average length of abstracts. In response, the *APA Publication Manual*, 5th edition, raised the word limit from 120 words to 200 words. The 6th edition specified a 250 word limit. We suggest flexibility on length.

In Checklist 11.1, we provide suggestions for what to include in a structured abstract. The items are consistent with those provided in the above-mentioned research on the value of structured abstracts, and

Checklist 11.1 Elements of a structured abstract

1. Purpose	What problem does the paper address?	☐
2. Methods	How was the problem addressed?	☐
3. Findings	What data were obtained, and what did the analysis show?	☐
4. Limitations	What are the caveats to the findings?	☐
5. Implications	What are the practical implications and why are they important?	☐

with the *Compliance With Science Checklist* (3.1). That said, journals might modify the structure in order to cater for the expectations and requirements of their own readers.

11.2.3 Require Disclosure, Justified Citations, Readability

Although required by virtually all of the leading scientific journals, full disclosure can be difficult to ascertain and so is not consistently enforced by journal editors. As a result, it is not possible to replicate many papers in scientific journals (see, e.g., McCullough, 2007).

In addition, when researchers plan to attempt a replication of a published study, they often find it necessary to contact the authors with questions about their data or methods, and they do not always get cooperation (Hubbard, 2016, p.149; Iqbal et al., 2016; McCullough et al., 2008). One study emailed the authors of 141 empirical papers that were published in the last two 2004 issues of four major APA journals asking for their data for the purpose of reanalysis. They had good reason to expect the cooperation of the authors because they would have been required to sign the *Certification of Compliance with APA Ethical Principles* that requires authors to make data available. After six months of repeated email requests, 73 percent of the authors had either refused or promised but failed to provide their data or had not responded the request (Wicherts et al., 2006).

The problem can be fixed by withholding publication until the information necessary for replication is provided. Resistance by some authors is likely, given the high likelihood that replications of observational studies would fail, and a tendency to secrecy by government agencies that have funded research (Cecil and Griffin, 1985). Nevertheless, some journals *do* enforce the requirement in that way, and doing so would seem likely to increase a journal's credibility and prestige.

Recognizing that for some papers it may not be possible to determine whether disclosure is complete or not without actually undertaking a replication, journals could include as a condition of publication that if disclosure is later found to be insufficient for replication, the paper will be retracted with a published explanation the authors put on a do-not-publish blacklist.

A more radical approach than "only" requiring disclosure of data and methods was proposed by Young and Karr (2011). Their 7-step process for observational studies requires researchers to specify models before obtaining the data, to then estimate models using only half of the data, and for the journal to publish an appendix with the analysis repeated for the second half of the data. Embarrassment is likely if the scientific method is not followed in developing the models and selecting and obtaining the data.

Some journals appear to be heading in that direction. For example, the journal publisher Wiley provides the options of "research pre-registration," and "registered reports." In early 2020, *PLoS ONE* announced that they would be accepting registered reports; essentially research paper proposals. *PLoS ONE* describes their Registered Report Protocols as requiring the rationale and methodology of the study, which would be published if the study meets *PLoS ONE*'s criteria. The publication of a Registered Report Research Article would follow the collection and analysis of data, provided that the conduct of the research followed the registered protocols.

Journals should require authors to sign an oath stating that all of the works cited in their paper were read by at least one of the authors and that the reason for citing each work is explained in the paper.

Finally, journals could increase the audience for papers by setting a readability criterion and standard for papers. If a paper is too complex, authors could be asked to have their paper copy edited. We typically use a number of copy editors for each of our publications.

11.2.4   Require an Oath that Standards Were Upheld

Much research has been done on cheating. Universities provide fertile ground for such studies given the incentives and the opportunities for cheating with which students and researchers are presented. Yet cheating – lying for personal gain or material payoffs – does not appear to be anywhere near as great as standard economic theory would predict.

Why not?

A meta-analysis of 90 experimental studies found that the best explanation for truth-telling against one's apparent interests was wanting to be honest and to be seen to be honest (Abeler et al., 2019). That desire may be due to social norms or standards that cause liars to suffer a non-material "lying cost."

Standards do affect behavior. And reminding people of an expected moral standard by requiring them to take an oath has a long history (see, e.g., Tyler, 1834).

Asking people to sign an honesty oath has been shown to reduce cheating. In one recent study, 29 subjects were asked to sign an oath that stated: "Hereby I do affirm that all the data I am about to provide regarding my actions during this experiment will be the absolute truth. I also do swear that all my actions during this experiment will be due to the principle of honesty and that I will not lie in order to enrich myself" (Beck et al., 2020, p. 480). Cheating among those subjects was assumed to be the reason for an average payoff that was 6 percent higher than the expected payoff without cheating, but their apparent cheating was 86 percent less than cheating among the 39 subjects who were not asked to sign an oath Group decision making, and monitoring treatments were ineffective or less effective in reducing cheating (Beck et al., 2020, p. 475., tab 2).

Scientists and others working on a scientific paper should be asked to sign an oath before starting a study. At that same time, they would be informed that they would also be expected to sign the oath when their research is completed.

The effect of that procedure can be enhanced by asking authors to include portrait photos of their face when they sign their oath or to do so while sitting in front of a mirror. The so-called mirror studies began in the 1970s – Silvia and Phillips (2013) include a summary of the development of objective self-awareness theory and experimental evidence on it. Authors might want to consider using their computer or phone camera to record themselves signing the oath.

11.2.5   Use Benford's Law to Identify Data Fabrication

Journals could announce in their instructions to authors that they use Benford's law, also known as the "first-digit rule," to check numerical data for fabrication (Berger and Hill, 2011). Benford's law is based on the observation that individual observations in a growing series or in a set of observations that encompass several orders of magnitude, the first digit of an observation is more likely to be 1 than any other digit, next most likely to be 2, and so on.

Journals could require the submission of the data, along with a statement in the instructions to authors that if their data fails Benford's law, the submitted paper would be rejected. The warning itself should be enough to discourage researchers from fabricating data.

Many studies have been published on Benford's law. In the United States, evidence based on Benford's law has been admitted in criminal cases at the federal, state, and local levels. Details about applying the law are also available on Wikipedia.

11.2.6   Insist on Cost–Benefit Implications of Effect Size Estimates

Journals could insist that papers include the information needed for cost–benefit analyses of alternative decisions or policies in response to the authors' findings of effect sizes. For example, a study that estimated a number of lives would be saved by reducing road speed limits suggests that the limits should be reduced, but analyzing all the costs and benefits might lead to a different conclusion. Providing such information would help readers to determine the importance of the findings for them and help them to make better decisions.

As we discussed in Chapter 6, Section 6.6 titled "Distracted by Statistical Significance Tests," the tests harm decision-making and have been rejected as evidence by the US Supreme Court (Ziliak, 2011). Researchers and reviewers have been advised to avoid relying on statistical significance test results by the American Statistical Association (Wasserstein and Lazar, 2016).

We suggest that any publication that claims to be a scientific journal should refuse to publish papers that include such tests.

11.2.7   Invite Papers from Leading Scientists

By invited, we mean that the paper will be accepted when the researcher decides it is ready. That eliminates the need for mandatory peer review. The *Journal of Economic Perspectives* has been successful in following that procedure. Their authors obtain reviews from eminent colleagues and they thank those reviewers in their papers.

An invitation is a contract to publish a paper without the requirement for mandatory peer review. Invited papers allow authors to choose topics that challenge current thinking, or to take on large tasks such as meta-analyses. It also allows authors to obtain their own reviews and to make the final decision on which suggestions to use.

In this book, as in our other research works, we have tried to use as many suggestions as we can that we believe result in improvements.

Invited papers help journals to publish papers that are more important than would otherwise be the case, and to do so less expensively, as the authors obtain reviews themselves. That strategy was used for the 1982 introduction of the *Journal of Forecasting*. Its impact factor for 1982 to 1983 was seventh among all journals in business, management, and planning.

An analysis of 545 papers on forecasting methods found that invited papers were rated as 20 times more important than those submitted in the traditional manner. The criteria were: (1) their findings were included in *Principles of Forecasting: A Handbook for Researchers and Practitioners*, a comprehensive collection of forecasting principles and (2) the number of citations the paper received (Armstrong and Pagell, 2003).

Another test on the value of invited papers was provided by an examination of the papers published in six leading economics journals from 1991 to 2000. Papers submitted by authors who had connections or personal ties to the editors of the journal in which they were published had more citations than did open submissions (Medoff, 2003).

Laband and Piette (1994) examined data from the social sciences on 1,051 papers published in 28 top economics journals in 1984. The average number of citations for papers by authors with connections to the editorial board was more than twice that for papers by authors with no connection.

11.2.8   Use Reviews to Improve, Not Reject, Papers

Here we are concerned with how journals could improve on current standard practice for reviewing papers. First of all, we suggest that journals instruct authors to obtain peer review prior to submitting their paper. Author-solicited peer reviews prior to journal submission help improve papers, as we discussed in Chapter 10.

Journals could request that authors provide a list of reviewers in their acknowledgments, and provide reviewers' contact information to the journal editor as a matter of record and for confirming that a review was provided if that seems appropriate.

Journal reviewers should not be asked for their recommendations on whether a paper should be published or not, but rather for suggestions on how the paper could be improved. Scott used that procedure when he was editor of the *Journal of Forecasting* and found that it yielded practical suggestions for improving papers at a low cost for processing the submission. The procedure seemed to be well-received by the authors of the papers. As a reviewer, Scott has for decades refrained from making recommendations about whether a paper should be selected for publication.

Frey (2003) concluded that journal editors are interested in publishing important papers. If editors are uncertain about whether a paper is important, they should ask colleagues, much as in the days prior to mandatory peer review. Reviewers, on the other hand, are concerned about whether there are any mistakes. There is a strong temptation to assume that a paper with findings that conflict with current beliefs must be mistaken.

The selection of an editor is important. Avoid appointing one who cannot believe that the conclusions could possibly be true and cannot imagine any evidence that would change that opinion.

Associate editors should be chosen on the basis that they provide independent opinions, much as happened in the early days of peer review. Some editors still take that approach. For example, Scott received a request from an associate editor of a journal who was concerned that reviewers had rejected a paper that he considered had merit. Scott quickly decided that the paper should be accepted for publication and recommended that it be published with commentary. It was likely rejected by the original reviewers because it challenged widely held beliefs. The paper – Soyer and Hogarth's (2012) descriptively and provocatively titled "The Illusion of Predictability: How Regression Statistics Mislead Experts" – went on to be widely cited and has been influential.

Journals could encourage ongoing moderated and open peer review on their journal's internet site, with responses from the authors. The reviews could be linked to the published paper to inform readers about any reviewer concerns with the research or alternative conclusions from the findings.

With open peer review on journal websites, we expect that reviews of important papers will pour in. The reviewers should be identified, and they should avoid opinions. Reviews should be linked to the reviewers and the authors should be free to publish on-line responses to the reviews.

Journals can encourage and publish all reviews posted by people who reveal their names, contact information, and potential biases. Reviewers should be directed to assess the evidence presented in the paper, and should avoid *ad hominem* arguments, and inflammatory opinions. Authors would be given advance notice of the reviews so that they have the opportunity to have their response published along with the review. Authors could request that their paper be withdrawn from the journal at any time if they came to believe that doing so was justified by the reviews.

For many years, the Internet has been used to get ratings about various products from many independent reviewers. Consider book reviews on the Amazon internet site. Those interested in the topic do the reviews; it is common that many reviews are published, they are submitted rapidly, and they are free. Some of them involve books about scientific issues. For example, John Lott's research findings on gun control in the third edition of his book *More Guns, Less Crime* (2010) had been reviewed by 76 readers on Amazon by March 2020. Fifty-seven reviews were positive and thirteen were critical.

Might inviting scientists to publish their papers, and to then request peer review from interested scientists provide sufficient quality control? The approach would encourage authors to withdraw their papers if they are not about to rebut claims of serious flaws in their study. The F1000 (Faculty of 1000) provides an example. It is currently used in over 400 disciplines, principally in the fields of biology and medicine.

11.2.9 Use the *Compliance With Science Checklist* to Structure Reviews

Journals could set their own standards for compliance with the Checklist 3.1 criteria, perhaps by specifying which of the criteria must be met in order to be accepted for publication. To that end, journals could include the *Compliance With Science Checklist* and the journal's criteria in their instructions for authors.

To use the *Compliance With Science Checklist* (Checklist 3.1) to obtain structured reviews of a paper, obtain reviews from three to five reviewers. The reviews can be completed, on average, in less than 30 minutes each.

Because they are assessing the use of the scientific method, reviewers do not need to be experts in the field of study. That suggests that the peer review should be done by paid reviewers, and the time from submission to publication of a paper could be much reduced. The cost could be borne by the authors. As we described in Chapter 5, the reliability of the procedure is high relative to reviews that are obtained using the currently standard journal peer review procedures.

Journal editors could provide the author with the raters' completed *Compliance With Science* checklists – explaining how their paper could be improved, so that the author can revise to improve compliance and resubmit. Authors might also be permitted to explain why some criteria do not apply to their paper, and such papers might be published along with the ratings and the authors' explanations.

As we discussed above, reviewers' *opinions* on whether to publish should not be requested, and should be ignored by editors if they are offered. That is the approach used by *PLoS ONE*. The decision of whether to publish should be a question only of whether the paper complies with the scientific method in ways and to the extent required by the journal and meets the journal's other, announced, requirements.

11.2.10   Use Operational Criteria to Select Objective Research Papers

Authors can face uncertainty about what criteria will be used to determine whether a paper will be accepted by a journal and, if it is accepted, how long they will have to wait for it to be published. So, what would happen if a journal removed uncertainty and provided operational criteria that if met would guarantee publication?

Rather than ask reviewers for their opinions on whether a *research* paper should be published in the journal, journals could specify objective criteria and enforce them. In addition to the journal's specific criteria – such as topic relevance – the requirement for full disclosure (Criterion 3 of the *Compliance With Science Checklist*) and an oath on standards as described earlier in this section, we suggest a criterion that if also met would qualify a paper for publication as a *research* paper: that is, objective testing.

We have discussed the importance of multiple reasonable hypothesis testing (MRHT) throughout this book as a way to achieve objectivity (Criterion 4 of the *Compliance With Science Checklist*). Insisting on MRHT would help journals to avoid publishing advocacy papers that would not bear the scrutiny of objective testing.

11.2.11 Create a Compliant with Science Papers Section

Journals could promise to publish all papers in the journal's field that comply to a sufficient extent with the *Compliance with Science Checklist* (Checklist 3.1) in a scientific papers section of the journal. To be considered for publication in the scientific papers section of the journal, the journal could require that authors include a completed and signed *Compliance With Science* Checklist with their paper submission.

Reviews of papers for compliance with science could be completed in a much more timely and objective way than is the case with traditional mandatory journal reviews. On receipt of a paper that is sufficiently compliant with the scientific method to meet the journal's declared standard on the basis of the authors' completed *Checklist*, an administrator should redact all potentially biasing information about the authors from the paper. Trained evaluators could then independently use the checklist to determine the degree to which the paper complies with science.

Rather than depend on reluctant and potentially biased volunteer reviewers, trained raters could be paid for reviews. Because the raters are asked only to assess the extent to which a paper follows the scientific method, it is not necessary to use reviewers who are experts on the topic of the paper, and so cost could be kept low by using, for example, Amazon's Mechanical Turk workers.

The *Compliance With Science Checklist* (Checklist 3.1) asks raters to describe the relevant aspect of a paper *in their own words* if they rate the paper as compliant against a checklist item. By using trained raters, rather than topic experts, as reviewers, journals can put the onus on authors to explain what they did and why the findings are important in a way that will be understood by an intelligent general reader.

A remarkable step in the direction of publishing all papers that comply with the scientific method in important ways was taken by *PLOS* (Public Library of Science), an online journal. To our knowledge, *PLOS* was the first scientific journal that presented a checklist to potential authors with the message that compliance will guarantee publication. It is based on seven required items, which are referred to as "soundness." They require that a paper: (1) provide results from primary research; (2) not be published elsewhere; (3) provide sufficiently detailed descriptions of analyses; (4) draw conclusions that are supported by the data; (5) be written in intelligible English; (6) meet ethical standards; and (7) adhere to standards for data availability (*PLoS ONE*, 2016a, 2016b). The papers are peer-reviewed with the aim of improving them.

Most important, however, is that *PLoS ONE's* stated policy is to eliminate "subjective assessments of significance or scope." Their policy addresses the problem of the bias against papers with important new findings. In addition, *PLoS ONE* has reduced the time needed to complete reviews and to make the papers available online. The journal reported a time-to-publication average of 170 days for the first half of 2021. In return for publishing papers that meet the soundness standard with Open Access, authors pay a fee.

Given the reductions in uncertainty for authors about criteria for acceptance, and the quick turnaround in publishing papers, it is not surprising that in the five years after its introduction in 2006, roughly 15,000 articles had been published in *PLoS ONE*. That made it the largest journal in the world at the time. Many *PLoS ONE* papers are important and widely read. The journal does well on the basis of citations, compared with established journals, and it appears to be a financial success (Straumsheim, 2017). *PLoS ONE* claimed an acceptance rate of nearly 50 percent for the first half of 2019, which is high relative to that of prestigious journals in the social and management sciences.

Following *PLoS ONE's* success, *Nature* launched an online open access journal in 2011, *Scientific Reports*, based on a similar model. Wikipedia listed twenty other "mega journals" that had as of May 2020 followed *PLoS ONE*. As far as we are aware, none has adopted criteria for publication that are as clear and rigorous in assessing scientific merit as those in the Table 3.1 *Compliance With Science Checklist*.

11.2.12 Publish Summaries of Useful Scientific Findings

The primary purpose of scientific journals is to publish and publicize scientific findings. To further that end, journals could provide an annual summary of useful scientific findings that they have published – providing the summaries in simple layman's terms would make it possible for all relevant stakeholders to benefit from the findings.

11.2.13 Continue Publishing Other Material of Interest to Scientists

None of the above suggestions requires that journals should eliminate sections for exploratory studies, theory development, applications, opinions, editorials, obituaries, tutorials, book reviews, commentaries, ethical issues, logical application of existing scientific knowledge, corrections, announcements, identification of problems in need of research, and so on. In short, journals should continue to provide a forum for cooperative efforts to improve science in a given field.

11.3   Governments, Regulators, and Courts

In Chapter 7 we examined evidence on the effect of government involvement in science. We describe how, even when motivated by wanting to make life better and safer for citizens, government regulation and funding of scientific research has the opposite effect.

With that evidence in mind, governments and their agencies should, when considering policies and regulations, rely on findings that are derived from studies that comply with the scientific method. Moreover, given that secrecy about data that was used in formulating and evaluating regulations is common (see Cecil and Griffin, 1985), we expect that most if not all current regulations should be reexamined for compliance with science.

Governments should not be involved in funding or doing research. One reason is that people elect politicians promising policies that they expect will benefit them while paying little regard to the cost to others, and without knowledge of long-term and unintended consequences. Governments will naturally commission advocacy studies to "sell" their policies to citizens.

If there is a demand for government action, we suggest that they issue public announcements of the problem that they are wanting to address and their interest in research of the likely effects of alternative policies and regulations – including the alternatives of making no change, and of deregulation – specifying that researchers must comply with *all of the criteria* for the scientific method (see Checklist 3.1, the *Compliance With Science Checklist*). In particular, the research should address all possible effects over the long run – both intended and unintended – so that a comprehensive cost–benefit analysis can be conducted.

There would be no need for governments to pay for such research, as it falls within the domains of interest of university research-ers, think tanks, and lobby groups of various kinds. The research reports and reviews of them, and completed *Compliance With Science Checklists* should be freely available in the public domain for all to see and to assess for themselves or via media summaries and commentaries.

Natural experiments provided by changes in government funding over time and between countries lead to a conclusion opposite to the popular misconception that people benefit when governments spend money on research (see e.g., Kealey, 1996). In practice, eco-nomic growth is slowed due the crowding out of private research and other more productive uses of resources. The history of governments' failures in "picking winners" and in advocating for causes that are unattractive to private investors is surely complete enough that no compelling argument can be mounted for government research funding and direction.

As we discuss throughout this book – and particularly in regard to universities in Section 11.1 of this chapter – free speech is vital to the advancement of scientific knowledge. Governments and courts can play a role in encouraging useful scientific contributions by protecting scien-tists' freedom of speech, and by otherwise staying out of the way.

Law courts can have a positive role by insisting that evidence is compliant with science. The Daubert standard of evidence from "scien-tific procedures" that was adopted after 1993 by the US Supreme Court – which we described in Chapter 1 – was a big step in that direction. As we discussed, however, there is a danger that the standard could end up replacing the previous standard of a consensus of scientists' opinions about the matter in contention with a consensus of scientists' opinions about which procedure should be used to settle the issue at stake. We suggest that if the evidence presented to a court were assessed using the *Compliance With Science Checklist*, courts would be able to make better informed judgments on the evidence presented to them.

11.4 Media and Interested Groups and Individuals

Reporting of claims of scientific findings should include easy access to the original study so that readers can judge for themselves what the paper concluded and how that was done. In recent years, the reporting of "science" in the mass media has been used for political purposes. It focuses primarily on the opinion of scientists who agree with a particular policy.

Media outlets should publish an ethics policy with details of how it will be enforced so that potential subscribers can judge which they will subscribe to. For example, Jim Lehrer, a popular journalist in the United States, described his personal principles for practicing journalism in the following 16 guidelines. We would hope that other journalists would agree with the principles.

1. Do nothing I cannot defend.
2. Do not distort, lie, slant, or hype.
3. Do not falsify facts or make up quotes.
4. Cover, write, and present every story with the care I would want if the story were about me.
5. Assume there is at least one other side or version to every story.
6. Assume the viewer is as smart and caring and good a person as I am.
7. Assume the same about all people on whom I report.
8. Assume everyone is innocent until proven guilty.
9. Assume personal lives are a private matter until a legitimate turn in the story mandates otherwise.
10. Carefully separate opinion and analysis from straight news stories and clearly label them as such.
11. Do not use anonymous sources or blind quotes except on rare and monumental occasions. No one should ever be allowed to attack another anonymously.
12. Do not broadcast profanity or the end result of violence unless it is an integral and necessary part of the story and/ or crucial to understanding the story.
13. Acknowledge that objectivity may be impossible but fairness never is.
14. Journalists who are reckless with facts and reputations should be disciplined by their employers.

15. My viewers have a right to know what principles guide my work and the process I use in their practice.
16. I am not in the entertainment business.

Jim Lehrer (1997, pp. 55–60)

Media and think tanks could also report on and keep running score cards on the extent to which governments' policies and regulations are compliant with science. They could do their own assessments guided by the *Compliance With Science Checklist*. Doing so would be greatly aided if the public had access to research reports along the line of those described in the previous section on governments' role in science.

As a final caution, journalists and interested others would benefit by keeping in mind Hayek's (1974) Nobel Prize speech warning about the divergence between what science can achieve and what the public, and some scientists, believe it can deliver:

> The conflict between what in its present mood the public expects science to achieve in satisfaction of popular hopes and what is really in its power is a serious matter because, even if the true scientists should all recognize the limitations of what they can do in the field of human affairs, so long as the public expects more there will always be some who will pretend, and perhaps honestly believe, that they can do more to meet popular demands than is really in their power. It is often difficult enough for the expert, and certainly in many instances impossible for the layman, to distinguish between legitimate and illegitimate claims advanced in the name of science. The enormous publicity recently given by the media to a report pronouncing in the name of science on The Limits to Growth, and the silence of the same media about the devastating criticism this report has received from the competent experts, must make one feel somewhat apprehensive about the use to which the prestige of science can be put.

12 RESCUING SCIENCE FROM ADVOCACY

The question remains: what ethical justification is there for imposing taxes on people to finance scientific research for which they would not voluntarily contribute? ...

Given the undisputed success of private financing in the past, the burden of proof that the benefits of government financing exceed the costs surely rests on those who support such financing. It is a burden they have hardly attempted to meet.

> Milton Friedman (quoting his 1981 "open letter" to the president of the National Academy of Sciences, 1994)

Researchers who have been applying the scientific method to important problems for over 20 centuries are responsible for saving lives and improving our quality of life. Their efforts have provided us with the comforts and the myriad of opportunities that we have to live fulfilling lives that could barely be imagined in earlier times.

Those efforts continue to save and improve lives, and there are still important problems that science can help us to solve, but current incentives and practices undermine scientific research. Research efforts are in many cases devoted to problems of little or no consequence and, even when the problems are important, researchers are failing to properly follow the scientific method, which can lead to useless or even harmful "solutions."

We set out to provide an operational definition of the scientific method that is based on the habits of thinking – including skepticism and objectivity – and the procedures – such as multiple reasonable hypothesis testing – described by leading figures of science.

In the private sector, science continues to be highly successful. Consider Silicon Valley. Privately funded experimental research in areas such as agriculture, computer technology, aeronautics, engineering, and medicine continues to produce useful scientific findings. Still, there are ways to improve practice in all areas. For example, an obvious change is to eliminate tests of statistical significance.

It seems safe to say that companies that use experimentation will grow faster than companies that do not or cannot experiment.

That is not the case in the public sector, however. Governments currently fund and regulate much research. That leads to what we have referred to as advocacy research: testing only a single, favored, hypothesis – most likely a hypothesis consistent with the regulatory preferences of government officials.

Our book reports on the reformation that began in the 1970s, when many scientists identified practices that ignored or violated the scientific method. The involvement of governments in science exacerbated concerns over malpractice, particularly with the rise of funding advocacy research related to public policy. In addition, universities use criteria for evaluation that are not related to useful scientific discoveries.

However, we believe that those who have the capabilities and interests can still become successful doing useful scientific research.

We also believe that mandated and monitored operational checklists can improve the success of scientists. We recommend that Checklist 3.1, Compliance with Science, should be used as part of the contract with scientists when they are hired.

The US government now tries to control how research must be done when it involves human subjects, despite a lack of evidence that independent scientists have harmed subjects. As we have noted, the only major cases of unethical treatment of subjects that we were able to find, such as with eugenics research, were government-sponsored studies.

12.1 From Where Should Funding Come?

Historically, private parties – such as wealthy farmers, successful industrialists, or family – have funded scientific research. Many discoveries have also come from self-funding. For example, Hedy Lamar was an acclaimed Hollywood actress who realized that she had the ability and the resources to do scientific research. She was primarily self-taught and worked in her spare time on inventions, such as an improved traffic stoplight. During WWII, she learned that radio-controlled torpedoes could be jammed and set out to solve that problem; she developed a frequency-hopping signal that could not be tracked or jammed, earning a patent for her invention in 1942.

12.2 Advocacy

Advocacy masquerading as scientific research violates the basic principle of objectivity. Famous scientists since Aristotle have warned against advocacy, yet many journals – particularly in the social sciences – currently rush to publish advocacy studies. Advocacy is especially common when the topic is controversial.

Government grants for research that support preferred policies are a major source of advocacy. They perpetuate the protection of "preferred" hypotheses over objective testing. Grants motivate researchers to work on problems even if those particular problems do not take full advantage of their knowledge and skills, and they undermine the intrinsic rewards of discovering useful scientific knowledge.

12.3 Bureaucracy and Irrelevant Incentives

An important cause of the deterioration in the practice of science since the 1970s is bureaucratization. People who are predisposed to be scientists are neither interested in working in groups whose members are selected to satisfy political objectives nor can they be expected to be properly motivated to work on solving problems of someone else's choosing.

Are there conditions under which bureaucracies are better than independent scientists at identifying important problems to be solved? We doubt it.

As a scientist and newspaperman, Benjamin Franklin understood that the role of scientists is to use the scientific method to make useful findings and to disseminate that important knowledge to the world. Measures such as publication counts, citation counts, and government grants are not relevant to those objectives. Thus, they are harmful to the advancement of science.

Private companies, too, can suffer from bureaucratic rules, as Chevassus-Au-Louis (2019) described. Scott worked for one such company, Eastman Kodak. After George Eastman died, the bureaucracy eroded objective scientific practice. For example, despite being the first to discover digital imaging, Kodak chose not to pursue digital photography because management thought it would harm profits from sales of film.

12.4   How Stakeholders Can Contribute

We provide suggestions for each stakeholder group on how they can help to improve the practice of science. It starts with scientists. They do not have to wait for permission to make changes.

12.4.1   Scientists

There will always be scientists. They will know who they are. They will usually start when young it seems, when they are teenagers, and will continue as long as they can, even without being paid.

Scientists have two primary objectives – to discover useful scientific findings and to disseminate that knowledge. They should be judged only upon their achievement of those objectives. Scientists themselves, and not regulators, should assure readers that they have met those objectives.

The definition of the scientific method has been consistent among renowned scientists for centuries, and we do not believe it requires any changes.

Creativity is required for scientific findings. Einstein excelled at that. That said, we conclude that useful scientific findings require properly designed experiments.

We offer checklists to aid scientists in complying with the scientific method. Other than operational, mandated, and monitored checklists, we are unaware of procedures that will lead to compliance with the scientific method.

We have known for centuries that, to guard against advocacy, we need to test multiple reasonable hypotheses.

Mistakes are common in science and will continue to occur. Thus, researchers should inform journals about errors in their papers and either correct the errors or retract the paper. Journals, in turn, should publicize the corrections and retractions so as to alert readers. Scientists should also quietly inform other researchers about possible mistakes in their work.

We recommend that scientists should not accept government funds because of the strings attached. It is difficult to believe that governments can identify important problems, suggest research that is relevant and cost-effective, and find researchers whose skills and interests match the nature of the problem.

12.4.2 Universities

Universities do not do scientific research. Scientists do that. The individual scientist knows what skills and interests he or she has. Scientists are best able to identify who would benefit from their research, who can provide useful peer reviews, and, if necessary, who might be willing to fund their research.

Universities should employ independent, creative, and objective scientists. To accomplish that, we suggest using Checklist 3.1 (Selecting Scientists to Make Useful Scientific Discoveries). They should also provide scientists with a creative and objective environment in which to conduct their research.

Universities should protect the freedom of scientists to select their own topics. They should also avoid suggesting research topics for scientists by offering grant money for specific projects. Finally, they should not "advise" researchers to gain approval for their research designs.

Universities should also help researchers disseminate their useful scientific findings. That involves making the research findings inexpensive and easy to obtain.

Most importantly, universities should protect scientists from attempts to censor them. To this end, many universities are adopting the University of Chicago free-speech protections (Zimmer et al., 2014, p. 1).

12.4.3 Media

Some news media and journalists have openly stated that they will not cover scientific papers or books that reach conclusions with which they disagree.

Journalists can use the *Compliance With Science Checklist* (Checklist 3.1) to inform themselves about the confidence one can have in scientific papers. For example, they could alert readers when findings in a paper are based on non-experimental data.

Journalists covering research should provide a free web link to the original paper. Readers can then reach their own conclusions. Many media already do this.

Media should resist efforts by governments to influence how they report science.

On the positive side, the Internet has empowered scientists to present their findings directly to many people.

12.4.4 Courts of Law

Courts of law can use the *Compliance With Science Checklist* to assess the integrity of scientific evidence. Unlike the Frye and the Daubert guidelines, no *expert* judgments are required. We expect that there is an advantage to using people who are not experts on the specific problem to assess the evidence as they may be less likely to be biased for or against a particular finding. We expect that raters could be hired and trained in about one hour by using a self-training module on the Internet.

The *Compliance With Science Checklist* (Checklist 3.1) can also be used by courts to help them to fulfill their role as a check-and-balance to legislators and regulators. Those affected by laws and regulations should expect the courts to strike them down if they are based on studies or assumptions that do not comply with science.

12.4.5 Separating Science from the State

We conclude that governments can best help scientific progress by (1) abstaining from funding non-defense research, (2) eliminating regulations that control scientists' research topics and designs, (2) eliminating direct involvement in any research, and (4) ensuring scientists' rights of free speech. In these conclusions, we are following the principle espoused by Thomas Jefferson (1779), by Milton Friedman (1981, 1994) quoted above, and by Richard Feynman who, in a 1963 lecture stated that "No government has the right to decide on the truth of scientific principles, nor to prescribe in any way the character of the questions investigated ... Instead it has a duty ... to maintain the freedom [of its citizens]" (Feynman 2015, p. 277).

Afterword

by Terence Kealey

As recently as 2005, the study of the scientific method centered on the writings of scholars such as Francis Bacon, René Descartes, Karl Popper, Paul Feyerabend, Imre Lakatos, and Thomas Kuhn, as was discussed by Alan Chalmers in his superb introductory book *What is This Thing Called Science?* (1976 [2013]).

But in 2005 John Ioannidis, then of the University of Ioannina in Greece, published his paper with the startling title of "Why Most Published Research Findings Are False," (2005b) and since then the study of the scientific method has centered on the writings of Daniele Fanelli, Brian Nosek, Paul Smaldino, and Richard McElreath, which have inspired this superb book by Scott Armstrong and Kesten Green.

In 2005 the names of the scholars changed because the emphasis in the question changed. The question always was: "How do we know this person's statement is true?" but before 2005 the emphasis came on the word "true" (as in "how do we know this person's statement is *true*?"), whereas today the emphasis has shifted to the word "this" (as in "how do we know *this* person's statement is true?").

The question of *this* person goes back to the founding in 1660 of the world's oldest surviving research society, the Royal Society of London, whose motto is *Nullius in verba*, "take nobody's word for it," by which the society has really meant, "take no single person's word for it." The society's founders recognized that people are fallible, so no single person's observations could be trusted: as

Thomas Sprat wrote in his official *History of the Royal Society of 1667*, individuals are biased by "pride, and the lofty conceit of men's own wisdom."

Lawyers, too, have long struggled with the question of "how do we know this person's statement is true?" so the early scientists decided to model their conventions on those of the common law: as Robert Boyle, the leading spirit of the early Royal Society, wrote in his *Sceptical Chymist* of 1661, scientists could copy lawyers in elaborating rules of evidence:

> Though the testimony of a single witness shall not suffice to prove the accused party guilty of murder; yet the testimony of two witnesses ... shall ordinarily suffice to prove a man guilty.

The Royal Society, therefore, in modelling itself on the law, moved to a collective model of research, performing experiments in public to achieve – in Sprat's words – the "unanimous advancement of the same works." But the philosopher Thomas Hobbes demurred.

Hobbes believed that no observations, however unanimous, could be accepted as truth, because even groups of individuals could be misled by collective errors. Only mathematics, Hobbes said, which was based on logic, could be unquestionably true. And because experiments were not unquestionably true, Hobbes said, the very activity of science was illegitimate. The Royal Society prided itself on the openness of its meetings, but eventually it was driven to exclude him from its premises.

Today, of course, scientists no longer perform experiments collectively and in public. Nor, when submitting papers from afar, do they copy early researchers like Antonie van Leeuwenhoek, who – when sending papers from his home in Delft, Holland, to the Royal Society in London – would also submit affidavits from prominent citizens confirming they'd seen under his microscopes what he'd claimed he'd seen. Instead, to lower transaction costs, scientists over the years were to move to a culture of trust, and to reinforce that trust they were to elaborate conventions such as transparent publication (i.e., each paper should contain sufficient information to allow duplication by others). That culture of trust was to serve scientists well until 2005, and it therefore allowed them the philosopher's luxury of asking "How do we know this person's statement is *true*?" rather than the lawyer's question of "how do we know *this* person's statement is true?"

Although that culture of trust was, in 2005 – on the publication of Ioannidis's paper – to be exposed as having failed, we can date the origins of the breakdown half a century earlier, back to 1950. Before 1950, scientific research in the United States and UK was funded privately, which worked superbly, as is witnessed by the Industrial Revolution having been first British, and then American; and as is further witnessed by the roll call of great Anglophonic scientists since Newton.

But in 1950 the Americans started to nationalize their science on the model of a country that did not forge the Industrial Revolution, and whose GDP *per capita* before 1945 had never, contrary to myth, achieved even 75 percent of Britain's (let alone have converged on the US's), yet which was formidable in war: Germany.[1] In his 1945 book *Science – the Endless Frontier* Vannevar Bush, a senior American science administrator, argued that future wars would be won only if the federal government funded pure science the way the Germans had long done; and in 1950, as the Cold War heated up, the federal government took Bush's advice when it created the National Science Foundation to fund peer-reviewed pure science across the nation's universities.

Bush's claim that superior advances in weapons technology would flow from federally-funded basic science were, however, to be discredited in 1969 by the Department of Defense, which in its *Project Hindsight* showed that – of 700 recent advances in weapons technology – only two had emerged out of the federal government's funding of basic science. Bush had been disproved: the two-decade old experiment in the federal government's funding of pure science had yielded little of the promised benefit.

But by 1969 the federal government in the United States had moved on from Germany, to remodel its science on the new cynosure: the USSR. In 1957 the Soviets had launched the first artificial satellite, *Sputnik*, which led many people to assume that socialist science yielded the greatest economic growth. So in 1959 and 1962 the two foundational papers in the modern economics of science were written by Richard Nelson and Kenneth Arrow respectively – Arrow was to win an economics Nobel prize, and though Nelson did not also win one, he

---

[1] For the references that disprove the myth of Germany overtaking the UK economically or technologically during the nineteenth century, please see Kealey, Terence; Martin Ricketts (April 6, 2020) *Innovative Economic Growth: Seven Stages of Understanding.* Washington, DC: The Cato Institute. www.cato.org/publications/economic-policy-brief/innovative-economic-growth-seven-stages-understanding.

was an economist of that calibre – to "prove" that governments had to fund science because science was best performed in a socialist setting.

The argument was of course false,[2] but it was celebrated by every vested interest: the universities loved it because it justified their funding by government; industry loved it because companies are always looking for corporate welfare; governments loved it because they could cast themselves as modern Medicis, sponsoring culture the way Ferdinand and Cosimo II in Florence had supported Galileo; and the general public loved it because, hey, what's not to love about planting the Stars and Stripes on the moon?

Since 1950 – and boosted by 1959 and 1962 – the governments of the world have therefore created a new ecosystem, namely the government-funded laboratory whose denizens answer only to peer-review. And, as we know from the science of epidemiology, if you create a new ecosystem, you also create a breeding ground for new diseases. Before 1950/1959/1962 universities had been essentially liberal arts colleges, funded by undergraduates and speaking truth unto power. Since then, though, they have become peer-reviewed research institutes, and they are now less interested in speaking truth unto power because said power is the source of their grants.

The universities took a few years to shift their focus, and during the 1950s Fred Stone of the National Institutes of Health described how "it wasn't anything to travel 200,000 miles a year" to persuade universities to submit grant applications. But the universities eventually changed, and students today are just a distraction from the professors' real goal, which is publishing peer-reviewed papers – which has bred a new disease: bias, in any and every shape and form, such that – in John Ioannidis's words – most published research findings are false.

I think only three people have identified the source of today's scientific distemper, namely Scott Armstrong and Kesten Green of this book, and Daniel Sarewitz of Arizona State University. In this book, Scott Armstrong and Kesten Green urge researchers to keep their research as practical as possible, for reasons Daniel Sarewitz summarized in his aphorism, "it's technology that keeps science honest." Before

---

[2] For an explanation of why today's widely-believed economics of science was always false, please see Kealey, Terence; Martin Ricketts (April 6, 2020) *Innovative Economic Growth: Seven Stages of Understanding.* Washington, DC: The Cato Institute. www.cato.org/publications/economic-policy-brief/innovative-economic-growth-seven-stages-understanding.

1950, most scientists were funded with technology in the funder's mind – which kept scientists honest. So, for example, if you're a scientist who is being funded by Thomas Edison to elaborate the laws of electricity to help his engineers develop better thermionic valves, and if the laws you elaborate are faulty, you won't find many hiding places in Menlo Park when it transpires the valves don't work because your science was biased.

But what if you're a university scientist who is funded by the review of your peers sitting in a committee far away in Washington, DC? And what if the peers who are reviewing your grants believe that fat and salt are bad for people? Or that there is no such thing as a safe dose of radiation or of fine particulates?[3] And yet you discover that the data do not support the opinions of the peer reviewers in Washington, DC; so what to do?

Do you do the honourable thing and try to publish papers showing that, actually, fat and salt are safe, and that low doses of radiation and fine particulates might be too? Or do you select the data that fit your peers' expectations and thus get the grants (and the papers and the promotions)? And do you persuade yourself you're doing the right thing? Do you, in fact, recreate in microcosm the dilemmas with which so many scientists struggled in Germany and the USSR, on whose systems of science the West has modelled itself?

In 2005 John Ioannidis told us what most scientists have done: he showed that most scientists select the data – and the statistics – to get the "right" answer. Not the truthful answer, the "right" one. And Daniele Fanelli, Brian Nosek, Paul Smaldino and Richard McElreath amongst others were to confirm he was correct. Before 1950, when most scientists were funded privately, their funding kept them honest because their science was audited when the thermionic valves hit the market; but in recent decades a scientist could aver that fat and salt are bad for people, or that there is no such thing as a safe dose of radiation or of fine particulates, and not only would such statements not damage a scientist's career, it would flourish *only* if he or she made those statements.

So, what to do? You'll find the answer in this book: be honest. And this book tells you *how* to be honest. Which, being the distillation

---

[3] See the examples in Patrick Michaels and Terence Kealey (2019), *Scientocracy* (pp. 139–160). Washington, DC: The Cato Institute.

of years of thought by Scott Armstrong and Kesten Green, tempered by the wisdom of the ages, is invaluable. Lots of people simply don't *know* how to be scientifically honest, but they do now because Scott Armstrong and Kesten Green have codified the wisdom of the last three and half centuries of scientific wresting with the question of "how do we know this person's statement is true?" So if you, the reader, follow this book's checklists and guides, you'll forge a career of which, ultimately, you'll be proud, even if – along the way – your road will be bumpier than that chosen by less scrupulous folk.

In the London *Spectator* of July 11, 2020 the historian David Wootton, reviewing Stuart Ritchie's *Science Fictions: Exposing Fraud, Bias, Negligence and Hype in Science*, suggested the problems in research had been recognized and were being corrected. I fear Wootton was being optimistic; rather, I think every PhD student should be provided by their supervisor (in British parlance) with a copy of Scott Armstrong's and Kesten Green's *Scientific Method*, to spark a full and frank discussion about the realities of life at the bench.

One of the beauties of science is its capacity to adapt to new realities. Since WWII, science has moved from being largely privately-funded to being largely publicly-funded, and in consequence its distempers have shifted. So, to meet those new distempers, Scott Armstrong and Kesten Green have proposed useful remedies. I think the spirit of Robert Boyle, looking down 350 years after he'd first struggled with the question of "how do we know this person's statement is true?" would be pleased with their book.

Terence Kealey lectured in clinical biochemistry at Cambridge between 1986 and 2000 and was the Vice-Chancellor (President) of the University of Buckingham – Britain's first independent university – between 2001 and 2014. He is currently an Adjunct Scholar at the Cato Institute in Washington, DC, and Emeritus Professor of Clinical Biochemistry at the University of Buckingham.

REFERENCES

Abeler, J., Nosenzo, D., & Raymond, C. (2019). Preferences for truth-telling. *Econometrica*, *87* (4), 1115–1153.

Abramowitz, S. I., Gomes, B., & Abramowitz, C. V. (1975). Publish or politic: Referee bias in manuscript review. *Journal of Applied Social Psychology*, *5* (3), 187–200.

Allen, M. (1991). Meta-analysis comparing the persuasiveness of one-sided and two-sided messages. *Western Journal of Speech Communication*, *55* (Fall), 390–404.

American Psychological Association (2001). *Publication Manual of the American Psychological Association*, (5th Ed.). American Psychological Association: Washington, D.C.

Arkes, H. R. (2003). The nonuse of psychological research at two federal agencies. *American Psychology Society*, *14* (1), 1–6.

Arkes, H. R., Gonzalez-Vallejo, C., Bonham, A. J., Kung, Y.-H., & Bailey N. (2010). Assessing the merits and faults of holistic and disaggregated judgments. *Journal of Behavioral Decision Making*, *23*, 250–270.

Arkes, H. R., Shaffer, V. A., & Dawes, R. M. (2006). Comparing holistic and disaggregated ratings in the evaluation of scientific presentations. *Journal of Behavioral Decision Making*, *19*, 429–439.

Armstrong, J. S. (1970), How to avoid exploratory research, *Journal of Advertising Research*, *10* (4), 27–30.

Armstrong, J. S. (1977). Social irresponsibility in management, *Journal of Business Research*, *5* (3), 185–213.

Armstrong, J. S. (1979). Advocacy and objectivity in science. *Management Science*, *25* (5), 423–428.

Armstrong, J. S. (1980a). The seer-sucker theory: The value of experts in forecasting, *Technology Review*, *83* (June/July), 16–24.

Armstrong, J. S. (1980b). Advocacy as a scientific strategy: The Mitroff myth. *Academy of Management Review*, *5* (4), 509–511.

Armstrong, J. S. (1980c). Unintelligible management research and academic prestige, *Interfaces, 10*, 80–86.

Armstrong, J. S. (1980d). Teacher vs. learner responsibility in management education. Working paper. Available at SSRN: http://dx.doi.org/10.2139/ssrn.647802.

Armstrong, J. S. (1982a). Barriers to scientific contributions: The author's formula. *Behavioral and Brain Sciences, 5*, 197–199.

Armstrong, J. S. (1982b). Is review by peers as fair as it appears? *Interfaces, 12* (5), 62–74.

Armstrong, J. S. (1983). Cheating in management science, *Interfaces, 13* (4), 20–29.

Armstrong, J. S. (1985). *Long-Range Forecasting: From Crystal Ball to Computer.* 2nd ed., New York, NY: John Wiley and Sons.

Armstrong, J. S. (1991). Prediction of consumer behavior by experts and novices. *Journal of Consumer Research, 18*, 251–256.

Armstrong, J. S. (1996). Management folklore and management science: On portfolio planning, escalation bias, and such. *Interfaces, 26* (4) 28–42.

Armstrong, J. S. (1997). Peer review for journals: Evidence on quality control, fairness, and innovation, *Science and Engineering Ethics, 3*, 63–84.

Armstrong, J. S. (2001a). *Evaluating forecasting methods. In Principles of Forecasting: A Handbook for Researchers and Practitioners.* New York, NY: Springer.

Armstrong, J. S. (2001b). *Principles of Forecasting: A Handbook for Researchers and Practitioners.* New York, NY: Springer.

Armstrong, J. S. (2001c). *Role playing: A method to forecast decisions. In Principles of Forecasting: A Handbook for Researchers and Practitioners.* New York, NY: Springer.

Armstrong, J. S. (2006). How to make better forecasts and decisions: Avoid face-to-face meetings, *Foresight: The International Journal of Applied Forecasting, 5* (Fall), 3–15.

Armstrong, J. S. (2007). Significance tests harm progress in forecasting, *International Journal of Forecasting, 23*, 321–327.

Armstrong, J. S. (2010). *Persuasive Advertising.* Basingstoke, UK: Palgrave Macmillan.

Armstrong, J. S. (2012a). Illusions in regression analysis. *International Journal of Forecasting, 28* (3), 689–694.

Armstrong, J. S. (2012b). Predicting job performance: The Moneyball factor, *Foresight, 25*, 31–34.

Armstrong, J. S. (2012c). Natural learning in higher education, in N. M. Seel, ed., *Encyclopedia of the Sciences of Learning*, pp. 2426–2433. New York, NY: Springer.

Armstrong, J. S., & Brodie, R. J. (1994). Effects of portfolio planning methods on decision making: Empirical results, *International Journal of Research in Marketing, 11*, 73–84.

Armstrong., J. S., & Green, K. C. (2012). Demand forecasting: Evidence-based methods. Posted on *Scholarly Commons* and other repositories.

Armstrong, J. S., & Green, K. C. (2018). Forecasting methods and principles: Evidence-based checklists. *Journal of Global Scholars of Marketing Science, 28*, 103–159.

Armstrong, J. S., & Green, K. C. (2019). Data models for forecasting: No reason to expect improved accuracy. *ResearchGate*, 1–4. DOI: 10.13140/ RG.2.2.10150.37441.

Armstrong, J. S., & Hubbard, R. (1991). Does the need for agreement among reviewers inhibit the publication of controversial findings? *Behavioral and Brain Sciences, 14*, 136–137.

Armstrong, J. S., & Overton, T. S. (1977). Estimating nonresponse bias in mail surveys. *Journal of Marketing Research, 14*, 396–402.

Armstrong, J. S., & Pagell, R. (2003). Reaping benefits from management research: Lessons from the forecasting principles project. *Interfaces, 33* (6), 91–111.

Armstrong, J. S., & Patnaik, S. (2009). Using quasi-experimental data to develop empirical generalizations for persuasive advertising. *Journal of Advertising Research, 49* (2), 170–175.

Armstrong, J. S., & Schultz, R. L. (1993). Principles involving marketing policies: An empirical assessment. *Marketing Letters, 4* (3), 253–265.

Armstrong, J. S., Brodie, R., & Parsons, A. (2001). Hypotheses in marketing science: Literature review and publication audit, *Marketing Letters, 12* (2), 171–187.

Armstrong, J. S., Coviello, N., & Safranek, B. (1993). Escalation bias: Does it extend to marketing? *Journal of the Academy of Marketing Science, 21* (3), 247–253.

Armstrong, J. S., Green, K. C., & Soon, W. (2008). Polar bear population forecasts: A public-policy forecasting audit. *Interfaces, 38*, 382–405.

Armstrong, J. S., Du, R., Green, K. C., et al. (2016). Predictive validity of evidence-based persuasion principles. *European Journal of Marketing, 50*, 276–293 (followed by Commentaries, pp. 294–316).

Arrow, K. J. (1962). The economic implications of learning by doing. *The Review of Economic Studies, 29*(3), 155–173. Available at www.jstor.org/stable/ 2295952

Asch, S. E. (1955). Opinions and social pressure. *Scientific American, 193* (5), 31–35.

Avorn, J. (2004). *Powerful Medicines: The Benefits, Risks and Costs of Prescription Drugs*. New York, NY: Alfred A. Knopf.

Bacon, F. (1620 [1863]). The New Organon: Or the True Directions Concerning the Interpretation of Nature. In *The Works of Francis Bacon* (Vol. VIII) being the Translations of the Philosophical Works Vol. I by J. Spedding, J. R. Ellis, & D. D.Heath, eds., Boston: Houghton Mifflin.

Baker, D., Lidster, K., Sottomayor, A., et al. (2014). Two years later: Journals are not yet enforcing the ARRIVE Guidelines on Reporting Standards for Pre-Clinical Animal Studies. *PLOS Biology, 12* (1): e1001756.

Baker, M. (2016). Is there a reproducibility crisis? A *Nature* survey lifts the lid on how researchers view the "crisis" rocking science and what they think will help. *Nature, 533* (26 May), 452–454.

Bandura, A. (2008). Michael J. Mahoney (1946–2006). *Constructivism in the Human Sciences, 12* (1&2), 31–33.

Barber, B. (1961). Resistance by scientists to scientific discovery, *Science, 134,* 596–602.

Barreira, P., Basilico, M. & Bolotnyy, V. (2018). Graduate student mental health: Lessons from American economics departments. *Harvard University Working Paper.*

Batson, C. D. (1975). Rational processing or rationalization? The effect of disconfirming information on a stated religious belief. *Journal of Personality and Social Psychology, 32* (1), 176–184.

Baxt, W. G., Waeckerie, J. F., Berlin, J. A., et al. (1998). Who reviews reviewers? Feasibility of using a fictitious manuscript to evaluate peer reviewer performance. *Annals of Emergency Medicine, 32* (3), 310–317.

Beall, J. (2012). Predatory publishers are corrupting open access. *Nature, 489,* 179.

Beaman, A. L. (1991). An empirical comparison of meta-analytic and traditional reviews. *Personality and Social Psychology Bulletin, 17* (3), 252–257.

Beardsley, M. C. (1950). *Practical Logic.* New York, NY. Prentice Hall.

Beck, T., Bühren, C., Frank, B., et al. (2020). Can honesty oaths, peer interaction, or monitoring mitigate lying? *Journal of Business Ethics, 163,* 467–484.

Bedeian, A. G., Taylor, S. G., & Miller, A. N. (2010). Management science on the credibility bubble: Cardinal sins and various misdemeanors. *Academy of Management Learning & Education, 9* (4), 715–725.

Begley, C. G., & Ellis, L. M. (2012). Raise standards for preclinical cancer research. *Nature, 483* (29 March), 531–533.

Ben-Shahar, O., & Schneider, C. E. (2014). *More Than You Wanted to Know: The Failure of Mandated Disclosure.* Princeton, NJ: Princeton University Press.

Berger, A., & Hill, T. P. (2011). A basic theory of Benford's Law. *Probability Surveys, 8,* 1–126.

Berger, J., & Pope, D. (2011). Can losing lead to wining? *Management Science, 57* (5), 817–827.

Berninger, M., Kiesel, F., Schiereck, D., et al. (2018). The Readability of Finance Articles and the Number of Citations: Can Articles Be Too Straightforward to Be Cited? Working Paper (Revised 10 Jan 2020). Available at SSRN. https://ssrn.com/abstract=3024694

Berschscheid, E., Baron, R. S., Dermer, M., et al. (1973). Anticipating informed consent: An empirical approach. *American Psychologist*, *28*, 913–925.

Beyer, J. M. (1978). Editorial policies and practices among leading journals in four scientific fields. *Sociological Quarterly*, *19*, 68–88.

Bijmolt, T. H. A., Heerde, H. J. van, & Pieters, R. G. M. (2005). New empirical generalizations on the determinants of price elasticity. *Journal of Marketing Research*, *42* (2), 141–156.

Bird, A. (2018). Thomas Kuhn. In Edward N. Zalta, ed., *The Stanford Encyclopedia of Philosophy*.

Björk, B., & Solomon, D. (2013), The publishing delay in scholarly peer-reviewed journals, *Journal of Informetrics*, *7* (4), 914–923.

Blass, T. (2009). *The Man Who Shocked the World*. New York, NY: Basic Books.

Boxer, A. (2020). *A Scheme of Heaven: Astrology and the Birth of Science*. London: Profile Books.

Boylan, J. B., Goodwin, P., Mohammadipour, M., et al. (2015). Reproducibility in forecasting research. *International Journal of Forecasting*, *31* (1), 79–90.

Boyle, R. (1661). *The Sceptical Chymist*. Crook: London, England. Available at https://books.google.com.au/books?hl=en&lr=&id=pxhKAQAAMAAJ&oi=fnd&pg=PP11&dq=Robert+Boyle+1661+sceptical+chymist&ots=iQwDaGcy-8&sig=gggy-ZuNfMVV3SfVdu_KT78mnL4&redir_esc=y#v=onepage&q=Robert%20Boyle%201661%20sceptical%20chymist&f=false

Bradley, J. V. (1981). Pernicious publication practices. *Bulletin of the Psychonomic Society*, *18* (1), 31–34.

Brembs, B., Button, K., & Munafo, M. (2013). Deep impact: Unintended consequences of journal rank. *Frontiers in Human Neuroscience*, *7*, 1–12.

Broad, W., & Wade, N. (1982). *Betrayers of the Truth: Fraud and Deceit in the Halls of Science*. New York, NY: Simon and Schuster.

Brush, S. G. (1977). The search for quality in university research publications. *Social Studies of Science*, *7*, 395–400.

Budd, J. M., Stevert, M., & Schultz, T. R. (1998). Phenomena of retraction: Reasons for retraction and citations to the publications. *Journal of the American Medical Association*, *280*, 296–297.

Burdick, A. (2017). "Paging Dr. Fraud": The fake publishers that are ruining science. The New Yorker, March 22.

Burnham, J. C. (1990). The evolution of editorial peer review. *JAMA*, *263* (10), 1323–1329.

Bush, V. (1945). *Science – The Endless Frontier*: A Report to the President by Vannevar Bush, Director of the Office of Scientific Research and Development, July 1945. Washington, DC: US Government Printing Office.

Calabrese, E. J. (2019). The troubled history of cancer risk assessment. *Regulation*, Spring, 16–19.

Campbell, D. T. (1979). Assessing the impact of planned social change. *Evaluation and Program Planning*, 2, 67–90.

Carlquist, S. (2009). Darwin on island plants. *Botanical Journal of the Linnean Society*, *161* (1), 20–25.

Carroll, A. E., & Doherty, T. S. (2019). Meat consumption and health: Food for thought. *Annals of Internal Medicine*, *171* (10), 767–768.

Ceci, S. J. (2020). Mistakes were made (but not by me). In R. J. Sternberg, ed., *My Biggest Research Mistake: Adventures and Misadventures in Psychological Research*. Thousand Oaks, CA: Sage, pp. 197–200.

Cecil, S., & Griffin, E. (1985). The role of legal policies in data sharing. In S. E. Fienberg, M. E. Martin, & M. L. Straf, eds., *Sharing Research Data* . Washington, DC: National Academy Press, pp. 148–198.

Center, F. J., & National Research Council (2011). *Reference Manual on Scientific Evidence*, 3rd ed. Washington, DC: National Academies Press.

Cerf, C., & Navasky, V. (1998). *The Experts Speak: The Definitive Compendium of Authoritative Misinformation*. New York, NY: Villard.

Chalmers, A. (2013). *What Is This Thing Called Science*, 4th ed. University of Queensland Press. Available at www.amazon.com.au/dp/B00CKD9ALK/ref= cm_sw_em_r_mt_dp_XZKB37KN1CXE4AMDYTA4

Chamberlin, E. H. (1948). An experimental imperfect market. *Journal of Political Economy*, *56* (2), 95–108. Available at www.jstor.org/stable/1826387

Chamberlin, T. C. (1890). The method of multiple working hypotheses. Reprinted in 1965 in *Science*, *148*, 754–759.

Chamberlin, T. C. (1899). Lord Kelvin's address on the age of the Earth as an abode fitted for life, *Science*, *9* (235), 889–901.

Chevassus-Au-Louis, N. (2019). *Fraud in the Lab: The High Stakes of Scientific Research*. Cambridge, MA: Harvard University Press.

Christensen-Szalanski, J. J. J., & Beach, L. R. (1984). The citation bias: Fad and fashion in the judgment and decision literature. *American Psychologist*, *39* (1), 75–78.

Cialdini, R. B. (2001). *Influence: Science and Practice*. Needham Heights, MA: Allyn and Bacon.

Cicchetti, D. V. (1991). The reliability of peer review for manuscript and grant submissions: A cross-disciplinary investigation. *Behavioral and Brain Sciences*, *14* (01), 119–135.

Clancy, D. (2017). A list of colleges that don't take Federal money. Dean Clancy Blog, December 2, updated August 10 2020. Available at https://deanclancy .com/a-list-of-colleges-that-donttake-federal-money/

Clinton, V. (2019). Reading from paper compared to screens: A systematic review and meta-analysis. *Journal of Research in Reading*, *42* (2), 288–325.

Cole, S. (1979). Age and scientific performance. *American Journal of Sociology*, *84* (4), 958–977.

Committee on Science, Space and Technology (2019). Hearing charter: Strengthening transparency or silencing science? The future of science in EPA rulemaking. US House of Representatives, November 13, 2318 Rayburn House Office Building.

CRASH trial collaborators (2004). Effect of intravenous corticosteroids on death within 14 days in 10 008 adults with clinically significant head injury (MRC CRASH trial): Randomised placebo-controlled trial. *The Lancet, 364* (9442), 1321–1328.

Crichton, M. (1975). Medical obfuscation: Structure and function. *New England Journal of Medicine, 293* (24), 1257–1259.

Czerlinski, J., Gigerenzer G., & Goldstein D. G. (1999). How good are simple heuristics? In Gerd Gigerenzer & Peter M. Todd, eds., *Simple Heuristics That Make Us Smart.* New York: Oxford University Press, pp. 97–118.

Davidoff, F., Batalden, P., Stevens, D., et al. for the SQUIRE development group (2008). Publication guidelines for quality improvement in health care: Evolution of the SQUIRE project. *Qual Safe Health Care, 17* (Suppl I), i3–i9.

Deci, E. L., Koestner, R., & Ryan, R. M. (1999). A meta-analytic review of experiments examining the effects of extrinsic rewards on intrinsic motivation. *Psychological Bulletin, 125* (6), 627–667.

Delgado, P., Vargas, C. Ackerman, R., et al. (2018). Don't throw away your printed books: A meta-analysis on the effects of reading media on reading comprehension. *Educational Research Review, 25,* 23–38.

Dewald, W. G., Thursby, J. G., & Anderson, R. G. (1986). Replication in empirical economics: The *Journal of Money, Credit and Banking Project. The American Economic Review, 76* (4), 587–603.

Diamond, J., & Robinson, J. A. (eds) (2010). *Natural Experiments of History.* Cambridge, MA: Harvard.

Dillman, D. A., Smyth, J. D., & Christian, L. M. (2014). *Internet, Phone, Mail, and Mixed-Mode Surveys:The Tailored Design Method,* 4th ed. New York, NY: Wiley.

DiNardo, J. (2018). Natural experiments and quasi-natural experiments. In *The New Palgrave Dictionary of Economics.* London: Palgrave Macmillan.

Dinges D. F. (1992). Adult napping and its effects on ability to function. In C. Stampi, ed., *Why We Nap.* Birkhäuser, Boston, MA, pp. 118–134.

Dockery, D. W., Pope, C. A., Xu, X., et al. (1993). An association between air pollution and mortality in six US cities. *New England Journal of Medicine, 329,* 1753–1759.

Doucouliagos, C., & Stanley, T. D. (2009). Publication selection bias in minimum-wage research? A meta-regression analysis. *British Journal of Industrial Relations, 47* (2), 406–428.

Duarte, J. L., Crawford, J. T., Stern, C., et al., (2015). Political diversity will improve social psychological science. *Behavioral and Brain Sciences, 38* (e130).

Duckworth, A. L., & Seligman, M. E. P. (2005). Self-discipline outdoes IQ in predicting academic performance of adolescents. *Psychological Science, 16* (12), 939–944.

Ducoin, Francis J., D.D.S., et al. v. Dr. Ana M. Viamonte Ros, in her official capacity as the State Surgeon General, et al. (2009). 2003 CA 696.

Dunning, T. (2012). *Natural Experiments in the Social Sciences: A Design-Based Approach.* Cambridge, UK: Cambridge University Press.

Dyson. F. J. (2008). *The Scientist as Rebel.* New York, NY: New York Review of Books.

The Economist (2016). Why research papers have so many authors. *The Economist, 421* (9017), Nov. 26, p. 76.

Eichorn, P., & Yankauer, A. (1987). Do authors check their references? A survey of accuracy of references in three public health journals. *American Journal of Public Health, 77* (8), 1011–1012.

Einhorn, H. J. (1972). Alchemy in the behavioral sciences. *Public Opinion Quarterly, 36* (3), 367–378.

Epstein, D. (2013). *The Sports Gene.* New York, NY: Current.

Eriksson, K. (2012). The nonsense math effect. *Judgment and Decision Making, 7* (6), 746–749.

Erlingsson, S. J. (2009). The Plymouth Laboratory and the institutionalization of experimental zoology in Britain in the 1920s. *Journal of the History of Biology, (42),* 151–183.

Evans, J. T., Nadjari, H. I., & Burchell, S. A. (1990). Quotational and reference accuracy in surgical journals: A continuing peer review problem. *JAMA, 263* (10), 1353–1354.

Evanschitzky, H., & Armstrong, J. S. (2013). Research with in-built replication: comment and further suggestions for replication research. *Journal of Business Research, 66* (9), 1406–1408.

Evanschitzky, H., Baumgarth, C., Hubbard, R., et al. (2007). Replication research's disturbing trend, *Journal of Business Research, 60* (4), 411–415.

Faigman, D. L. (2013). The Daubert revolution and the birth of modernity: managing scientific evidence in the age of science. *University of California Davis Law Review, 46,* 893–930.

Fanelli, D. (2009). How many scientists fabricate and falsify research? A systematic review and meta-analysis of survey data. *PLoS ONE 4* e5738.

Fang, F. C., Steen, R. G., & Casadevall, A. (2012). Misconduct accounts for the majority of retracted scientific publications. *PNAS, 109* (42), 17028–17033.

Feist, G. J. (1998). A meta-analysis of personality in scientific and artistic creativity. *Personality and Social Psychology Review, 2* (4), 290–309.

Feist, G. J. (2006). How development and personality influence scientific thought, interest, and achievement. *Review of General Psychology, 10* (2), 163–182.

Festinger, L., Rieken, H. W., & Schachter, S. (1956). *When Prophecy Fails. A Social and Psychological Study of a Modern Group that Predicted the Destruction of the World.* Minneapolis, MN: University of Minnesota Press.

Feynman, R. P. (2015). *The Quotable Feynman*, ed. M. Feynman. Princeton, NJ: Princeton University Press.

Fidler, F., & Wilcox, J. (2018). Reproducibility of Scientific Results. In Edward N. Zalta, ed., *The Stanford Encyclopedia of Philosophy* (Winter). Available at https://plato.stanford.edu/archives/win2018/entries/scientific-reproducibility/

Fire, M., & Guestrin, C. (2019). Over-optimization of academic publishing metrics: Observing Goodhart's Law in action. *GigaScience, 8* (6), 1–20.

Flyvbjerg, B. (2016). The fallacy of beneficial ignorance: A test of Hirschman's hiding hand. *World Development, 84,* 176–189.

Francis, B. (1986). The grievance industry: The Human Rights Commission presiding over Australia's fastest-growth sector. *Quadrant,* January/February, 102–104.

Franco, M., Orduñez, P., Caballero, B., et al. (2007). Impact of energy intake, physical activity, and population-wide weight loss on cardiovascular disease and diabetes mortality in Cuba, 1980–2005. *American Journal of Epidemiology, 166* (12), 1374–1380.

Franklin, B. (1743). A proposal for promoting useful knowledge. *Founders Online, National Archives* (http://founders.archives.gov/documents/Franklin/01-02-02-0092 [last update: 2016-03-28]). Source: The Papers of Benjamin *Franklin*, vol. 2, January 1, 1735, through December 31, 1744, ed. L. W. Labaree. New Haven, CT: Yale University Press, 1961, pp. 378–383.

Frederick, S. (2005). Cognitive reflection and decision making. *Journal of Economic Perspectives, 19* (4), 25–42.

Freedman, D. A. (1991). Statistical models and shoe leather. *Sociological Methodology, 21,* 291–313.

Frey, B. S. (2003). Publishing as prostitution: Choosing between one's own ideas and academic failure. *Public Choice, 116,* 205–223.

Frey, B. S. (2010). *Happiness: A Revolution in Economics.* Cambridge, MA: MIT Press.

Frey, B. S. (2018). *Economics of Happiness.* Cham, Switzerland: Springer.

Friedman, M. (1953). The methodology of positive economics, from Essays in Positive Economics, reprinted in D. M. Hausman, ed., The Philosophy of Economics: An Anthology, 3rd ed. Cambridge: Cambridge University Press, pp. 145–178.

Friedman, M. (1981) [1994]. Correspondence: National Science Foundation grants for economics. *Journal of Economic Perspectives, 8* (1), 199–200.

Friedman, M., & Schwartz, A. J. (1991). Alternative approaches to analyzing economic data. *American Economic Review, 81* (1), Appendix 48–49.

Gagne, M., & Deci, E. L. (2005). Self-determination theory and work motivation. *Journal of Organizational Behavior, 26,* 331–362.

Gal, D. (2006). A psychological law of inertia and the illusion of loss aversion. *Judgment and Decision Making, 1* (1), 23–32.

Gal, D., & Rucker, D. D. (2018a). The loss of loss aversion: Will it loom larger than its gain? *Journal of Consumer Psychology, 28* (3), 497–516.

Gal, D., & Rucker, D. D. (2018b). Loss aversion, intellectual inertia, and a call for a more contrarian science: A reply to Simonson & Kivetz and Higgins & Liberman. *Journal of Consumer Psychology, 28* (3), 533–539.

Galton, F. (1872). Statistical inquiries into the efficacy of prayer. *Fortnightly Review, LXVIII* August 1, 125–135.

Galton, F. (1874). *English Men of Science: Their Nature and Nurture.* London: Macmillan.

Gans, J. S., & Shepherd, G. B. (1994). How are the mighty fallen: Rejected classic articles by leading economists. *Journal of Economic Perspectives, 8* (1), 165–179.

Gao, J., & Zhou, T. (2017). Stamp out fake peer review. *Nature, 546,* 33.

Garvey, W. D., Lin, N., & Nelson, C. E. (1970). Communication in the physical and the social sciences. *Science, 170* (3963), 1166–1173.

Gelles, D. (2018). James Dyson: "The public wants to buy strange things." *New York Times,* December 5.

Gigerenzer, G. (1991). How to make cognitive illusions disappear: Beyond "heuristics and bias." *European Review of Social Psychology, 2,* 83–115.

Gigerenzer, G. (2000). *Adaptive Thinking: Rationality in the Real World.* New York, NY: Oxford University Press.

Gigerenzer, G. (2015). On the supposed evidence for libertarian paternalism. *Review of Philosophy and Psychology, 6,* 361–383.

Gigerenzer, G. (2018a). Statistical rituals: The replication delusion and how we got there. *Advances in Methods and Practices in Psychological Science, 1* (2), 198–218.

Gigerenzer, G. (2018b). The bias bias in behavioral economics. *Review of Behavioral Economics, 5* (3–4), 303–336.

Gigerenzer, G., & Todd, P. M. (2000). *Simple Heuristics That Make Us Smart.* New York, NY: Oxford.

Gigerenzer, G., Czerlinski, J., & Martignon, L. (1999). In J. Shanteau, B. Mellers, & D. A. Schum, eds., *Decision Science and Technology.* Boston, MA: Springer, pp. 81–103.

Gigerenzer, G., Krauss, S., & Vitouch, O. (2004). The null ritual: What you always wanted to know about significance testing but were afraid to ask. In

D. Kapli, ed., *The Sage Handbook of Quantitative Methodology for the Social Sciences.* Thousand Oaks, CA: Sage, pp. 391–408.

Goodstein, L. D., & Brazis, K. L. (1970). Psychology of scientist: XXX. Credibility of psychologists: An empirical study. *Psychological Reports, 27* (3), 835–838.

Gordon, G., & Marquis S. (1966). Freedom, visibility of consequences, and scientific innovation. *American Journal of Sociology, 72* (2), 195–202.

Gould, S. J. (1970). Evolutionary paleontology and the science of form. *Earth-Science Reviews, 6,* 77–119.

Green, K. C. (2002). Forecasting decisions in conflict situations: A comparison of game theory, role-playing, and unaided judgement. *International Journal of Forecasting, 18,* 321–344.

Green, K. C. (2005). Game theory, simulated interaction, and unaided judgement for forecasting decisions in conflicts: Further evidence. *International Journal of Forecasting, 21,* 463–472.

Green, K. C., & Armstrong, J. S. (2007a). The value of expertise for forecasting decisions in conflicts. *Interfaces, 37* (3), 287–299.

Green, K. C., & Armstrong, J. S. (2007b). Structured analogies for forecasting. *International Journal of Forecasting, 23,* 365–376.

Green, K. C., & Armstrong, J. S. (2008). Uncertainty, the precautionary principle, and climate change. Available at www.researchgate.net/publication/348663804_Uncertainty_the_Precautionary_Principle_and_Climate_Change

Green, K. C., & Armstrong J. S. (2012). Evidence on the effects of mandatory disclaimers in advertising. *Journal of Public Policy and Marketing, 31* (2), 325–325.

Green, K. C., & Armstrong, J. S. (2015). Simple versus complex forecasting: The evidence. *Journal of Business Research, 68,* 1678–1685.

Grove, W. M., Zald, D. H., Lebow, B. S., et al. (2000). Clinical versus mechanical prediction: A meta-analysis, *Psychological Assessment, 12* (1), 19–30.

Guyatt, G. H., Oxman, A. D., Vist, G. E., et al. (2008). GRADE: An emerging consensus on rating quality of evidence and strength of recommendations. *British Medical Journal, 336,* 924–926.

Hair, K., Macleod, M. R., Sena, E. S., & The IICARus Collaboration (2018). A randomised controlled trial of an intervention to improve compliance with the ARRIVE guidelines (IICARus).

Hales, B. M., & Pronovost, P. J. (2006). The checklist – a tool for error management and performance improvement. *Journal of Critical Care, 21,* 231–235.

Hangarter, R. P. (2000). *Darwin and His Research on Plant Motion.* Plants-In-Motion, Indiana University website.

Hardin, G. (1968). The tragedy of the commons. *Science, 162* (3859), 1243–1248. DOI: 10.1126/science.162.3859.1243.

Hartley, J. (2003). Improving the clarity of journal abstracts in psychology: A case for structure. *Science Commnication, 24* (3), 366–379.

Hauer, E. (2004). The harm done by tests of statistical significance. *Accident Analysis and Prevention, 36* (3), 495–500.

Hauer, E. (2019). On the relationship between road safety research and the practice of road design and operation. *Accident Analysis and Prevention, 128,* 114–131.

Hayek, F. A. von (1974). Prize lecture: The pretence of knowledge. *NobelPrize. org,* Lecture to the memory of Alfred Nobel, December 11, 1974.

Haynes, A. B., Weiser, T. G., Berry, W. R., et al. (2009). A surgical safety checklist to reduce morbidity and mortality in a global population. *New England Journal of Medicine, 360* (5), 491–499.

Hertwig, R., & Ortmann, A. (2008). Deception in experiments: revising the arguments in its defense. *Ethics & Behavior, 18* (1), 59–92.

Higgins, E. T., & Liberman, N. (2018). The loss of loss aversion: Paying attention to reference points. *Journal of Consumer Psychology, 28* (3), 523–532.

Hirschman, A. O. (1967). The principle of the hiding hand. *The Public Interest, 6* (Winter), 1–23.

Hogarth, R. M. (2012). When simple is hard to accept. In P. M. Todd & G. Gigerenzer, eds., & ABC Research Group, *Evolution and Cognition. Ecological Rationality: Intelligence in the World.* New York, NY: Oxford University Press, pp. 61–79.

Hollingworth, H. L. (1913). *Advertising and Selling: Principles of Appeal and Response.* New York: D. Appleton & Co.

Holub H. W., Tappeiner, G., & Eberharter, V. (1991). The iron law of important articles. *Southern Economic Journal, 58,* 317–328.

Horwitz, S. K., & Horwitz, I. B. (2007). The effects of team diversity on team outcomes: A meta-analytic review of team demography. *Journal of Management, 33* (6), 987–1015.

Hubbard, R. (2016). *Corrupt Research: The Case for Reconceptualizing Empirical Management and Social Science.* New York, NY: Sage.

Hubbard, R., & Armstrong, J. S. (1994). Replications and extensions in marketing: Rarely published but quite contrary. *International Journal of Research in Marketing, 11,* 233–248.

Hubbard, R., & Vetter, D. E. (1996). An empirical comparison of published replication research in accounting, economics, finance, management, and marketing. *Journal of Business Research, 35* (2), 153–164.

Hubbard, R., Vetter, D. E., & Little, E. L. (1998). Replication in strategic management: Scientific testing for validity, generalizability, and usefulness. *Strategic Management Journal, 19* (3), 243–254.

Idaho Statesman (1901). Doing one's best. *Idaho Daily Statesman*, May 6, p. 4, col. 3, Boise Idaho. https://quoteinvestigator.com/2012/12/14/genius-ratio/#note-5018-8

Inbar, Y., & Lammers, J. (2012). Political diversity in social and personality psychology. *Perspectives on Psychological Science*, 7 (5), 496–503. DOI: 10.1177/1745691612448792.

Infectious Diseases Society of America (2009). Grinding to a halt: The effects of the increasing regulatory burden on research and quality. *Clinical Infectious Diseases*, 49 (1), 328–335.

Ioannidis, J. P. A. (2005a). Why most published findings are false. *PLOS Medicine*, 2(8): e124. DOI: 10.1371/journal.pmed.0020124.

Ioannidis, J. P. A. (2005b). Contradicted and initially stronger effects in highly cited clinical research. *JAMA*, *294*, 218–228

Ioannidis, J. P. A. (2014). Is your most cited work your best? *Nature*, *514*, 561–562.

Iqbal, S. A., Wallach, J. D., Khoury, M. J., et al. (2016). Reproducible research practices and transparency across the biomedical literature. *PLOS Biology*, *14* (1). DOI:10.1371/journal.pbio.1002333

Iyengar, S. S., & Lepper, M. R. (2000). When choice is demotivating: Can one desire too much of a good thing? *Journal of Personality and Social Psychology*, *79* (6), 995–1006.

Jacquart, P., & Armstrong, J. S. (2013). Are top executives paid enough? An evidence-based review. *Interfaces*, *43* (6), 580–589.

Jamali, H. R. (2017). Copyright compliance and infringement in ResearchGate full-text journal articles. *Scientometrics*, *112*, 241–254.

Jauch, L. R., & Wall, J. L. (1989). What they do when they get your manuscript: A survey of *Academy of Management* reviewer practices. *Academy of Management Journal*, *32* (1), 157–173.

Jefferson, T. (1779). 82. A bill for establishing religious freedom, 18 June, 1779. *Founders Online*, National Archives, accessed June 3, 2019, https://founders.archives.gov/documents/Jefferson/01-02-02-0132-0004-0082.

Johansen M., & Thomsen S. F. (2016). Guidelines for reporting medical research: A critical appraisal. *International Scholarly Research Notices*, *2016*, 1–7.

John, L. K., Lowenstein, G., & Prelec, D. (2012). Measuring the prevalence of questionable research practices with incentives for truth telling. *Psychological Science*, *23* (5), 524–532.

Kabat, G. C. (2008). *Hyping Health Risks*. New York, NY: Columbia University Press.

Kahneman, D. (2011). *Thinking, Fast and Slow*. New York, NY: Farrar, Straus and Giroux.

Kahneman, D., & Tversky, A. (1979). Prospect theory: An analysis of decision under risk. *Econometrica*, *47* (2), 263–292.

Karau, S. J., & Williams, K. D. (1993). Social loafing: A meta-analytic review and theoretical integration. *Journal of Social Psychology*, *65* (4), 681–706.

Karpoff, J. M. (2001). Private versus public initiative in Arctic exploration: The effects of incentives and organizational structure. *Journal of Political Economy, 107* (4), 38–78.

Kealey, T. (1996). *The Economic Laws of Scientific Research*. London: Macmillan.

Kendall, P. C., & Ford, J. D. (1979). Reasons for clinical research: Characteristics of contributors and their contributions to the Journal of Consulting and Clinical Psychology. *Journal of Consulting and Clinical Psychology, 47* (1), 99–105.

Kerr, N. L., & Tindale, R. S. (2004). Group performance and decision making. *Annual Review of Psychology, 55*, 623–655.

King, D. W., McDonald, D. D., & Roderer, N. K. (1981). *Scientific Journals in the United States: Their Production, Use and Economics*. New York, NY: Hutchinson and Ross.

Koehler J. J. (1993). The influence of prior beliefs on scientific judgments of evidence quality. *Organizational Behavior and Human Decision Processes, 56* (1), 28–55.

Koning, A. J., Franses, P. H., Hibon, M., & Stekler, H. O. (2005). The M3 competition: Statistical tests of the results. *International Journal of Forecasting, 21* (3), 397–409.

Krugman, H. E. (1965). The impact of television advertising: Learning without involvement. *Public Opinion Quarterly, 299*, 349–356.

Kuenen, P. H. (1958). Experiments in Geology. *Transactions of the Geological Society of Glasgow, 23*, 1–28.

Kühberger, A. (1998). The influence of framing on risky decisions: A meta-analysis. *Organizational Behavior and Human Decision Processes, 75* (1), 23–55.

Kuhn, T. S. (1962). *The Structure of Scientific Revolutions*. Chicago, IL: University of Chicago Press.

Kupfersmid, J., & Wonderly, D. M. (1994). *An Author's Guide to Publishing Better Articles in Better Journals in the Behavioral Sciences*. Hoboken, NJ: Wiley.

Laband, D. N., & Piette, M. J. (1994). Favoritism versus search for good papers: Empirical evidence regarding the behavior of journal editors. *Journal of Political Economy, 102* (1), 194–203.

Laframboise, D. (2020a). Cancel culture hits medical journals. *BigPicNews.com*, January 27.

Laframboise, D. (2020b). Here's who pressured the medical journal. *BigPicNews.com*, January 29.

Langbert, M., Quain, A. J., & Klein, D. B. (2016). Faculty voter registration in economics, history, journalism, law, and psychology. *Econ Journal Watch, 13* (3), 422–451.

Latham, G. P., & Locke, E. A. (1979). Goal setting – a motivational technique that works. *Organizational Dynamics, 8* (2), 68–80.

Lau, R. D. (1994). An analysis of the accuracy of "trial heat" polls during the 1992 presidential election. *Public Opinion Quarterly, 58* (1), 2–20.

Lehrer, J. (1997). Individual statements of core journalistic values. In C. M. Firestone, ed., *The 1997 Catto Report on Journalism and Society.* Washington, DC: The Aspen Institute.

Levin, I. P., Schneider, S. L. & Gaeth, G. J. (1998). All frames are not created equal: A typology and critical analysis of framing effects. *Organizational Behavior and Human Decision Processes, 76* (2), 149–188.

Lewis, M. (2003). *Moneyball.* New York, NY: Norton.

Lewis, S. (1925). *Arrowsmith.* Harcourt Brace.

Lindsay, J. A., Boghossian, P., & Pluckrose, H. (2018). Academic grievance studies and the corruption of scholarship. *Areo,* October 2.

Lindsey, D. (1978). *The Scientific Publication System in Social Science.* San Francisco, CA: Jossey-Bass.

Lock, S., & Smith, J. (1986). Peer review at work. *Journal of Scholarly Publishing, 17* (4), 303–316.

Lock, S., & Smith, J. (1990). What do peer reviewers do? *JAMA, 263* (10), 1341–1343.

Locke, E. A. (1986). *Generalizing from Laboratory to Field Settings.* Lexington, MA: Lexington Books.

Locke, E. A., & Latham, G. P. (2019). Does prospect theory add or subtract from our understanding of goal directed motivation? In D. L. Sone & J. H. Dulebohn, eds., *The Only Constant in HRM Today Is Change.* Charlotte, NC: Information Age, pp. 19–41.

Locke, E. A., Schattke, K. (2019). Intrinsic and extrinsic motivation: Time for expansion and clarification. *Motivation Science, 5* (4), 277–290.

Lott, J. (2010). *More Guns, Less Crime.* Chicago, IL: University of Chicago Press.

Lott, M. (2014). Over 100 published science journal articles just gibberish. *FoxNews.com,* March 01.

Lovato, N., & Lack, L. (2010). The effects of napping on cognitive functioning. *Progress in Brain Research, 185,* 155–166.

Lu, S. F., Jin, G. Z., Uzzi, B., Jones, B. (2013). The retraction penalty: Evidence from the web of science. *Scientific Reports, 3* (3146), 1–5.

MacGregor, D. G. (2001). Decomposition for judgmental forecasting and estimation. In J. S. Armstrong, ed., *Principles of Forecasting.* London: Kluwer Academic Publishers, pp. 107–123.

MacKay, C. (1841). *Memoirs of Extraordinary Popular Delusions & the Madness of Crowds.* New York, NY: Three Rivers Press.

McCloskey, D. N., & Ziliak, S. T. (1996). The standard error of regressions. *Journal of Economic Literature, 34* (March), 97–114.

McCord, J. (1978). A thirty-year follow-up of treatment effects, *American Psychologist, 33* (3) 284–290.

McCullough, B. D. (2000). Is it safe to assume that software is accurate? *International Journal of Forecasting, 16*, 349–357.

McCullough, B. D. (2007). Got replicability? The *Journal of Money, Credit and Banking* Archive. *Econ Journal Watch, 4(3)*, 326–337.

McCullough, B. D., McGeary, K. A., & Harrison , T. G. (2008). Do economics journal archives promote replicable research? *Canadian Journal of Economics, 41*, 1406–1420.

McShane, B. B., & Gal, D. (2015). Blinding us to the obvious? The effect of statistical training on the evaluation of evidence. *Management Science, 62* (6), 1707–1718.

McShane, B. B., & Gal, D. (2017). Statistical significance and the dichotomism of evidence. *Journal of the American Statistical Association, 112* (519), 885–895.

Mahoney, M. J. (1976). *Scientist as Subject: The Psychological Imperative.* Cambridge, MA: Ballinger.

Mahoney, M. J. (1977). Publication prejudices: An experimental study of confirmatory bias in the peer review system. *Cognitive Therapy and Research, 1* (2), 161–175.

Mahoney, M. J., & DeMonbreun, B. G. (1977). Psychology of the scientist: An analysis of problem-solving bias. *Cognitive Therapy and Research, 1* (3), 229–255 (with commentaries).

Mahoney, M. J., & Kimper, T.P. (1976). *From ethics to logic: A survey of scientists.* In M. Mahoney, ed., *Scientist as Subject: The Psychological Imperative,* Cambridge, MA: Ballinger, pp. 187–194.

Maier, N. R. F. (1963). *Problem-Solving Discussions and Conferences: Leadership Methods and Skills.* New Your, NY: McGraw-Hill.

Maier, N. R. F., & Hoffman, L. R. (1960). Quality of first and second solutions in group problem solving. *Journal of Applied Psychology, 44* (4), 278–283.

Mansfield, E. (1980). Basic research and productivity increase in manufacturing. *The American Economic Review, 70* (5), 863–873.

Mansfield, E. (1998). Academic research and industrial innovation: An update of empirical findings. *Research Policy, 26*, 773–776.

Marquis, M. J., Warren, E. S., & Arnkoff, D. (2009). Michael J. Mahoney: A retrospective. *Journal of Psychotherapy Integration, 19* (4), 402–418.

Martin, B. R. (2013). Whither research integrity? Plagiarism, self-plagiarism and coercive citation in an age of research assessment. *Research Policy, 42* (5), 1005–1014.

Mayo, D. G. (1996). *Error and the Growth of Experimental Knowledge.* Chicago, IL: University of Chicago Press.

Mayr, E. (1997). *This Is Biology: The Science of the Living World.* Cambridge, MA: Harvard.

Medoff, M. H. (2003). Editorial favoritism in economics? *Southern Economic Journal*, 70 (2), 425–434.

Medvedev, Z. A. (1969). *The Rise and Fall of T. D. Lysenko*. New York, NY: Columbia.

Meehl, P. E. (1954). *Clinical Versus Statistical Prediction*: A Theoretical Analysis and a Review of the Evidence. Minneapolis, MN: University of Minnesota Press.

Meehl, P. E. (1978). Theoretical risks and tabular asterisks: Sir Karl, Sir Ronald, and the slow progress of soft psychology. *Journal of Consulting and Clinical Psychology*, 46, 806–834.

Milgram, S. (1963). Behavioral study of obedience. *Journal of Abnormal Psychology*, 67 (4), 371–378.

Milgram, S. (1974). *Obedience to Authority*. New York: Harper & Row.

Miller, D. W., Jr. (2007). The government grant system: Inhibitor of truth and innovation? *Journal of Information Ethics*, Spring, 59–69.

Miller, H. I. (1996). When politics drives science: Gore, and US Biotechnology Policy. *Social Philosophy and Policy*, 13 (2), 96–112.

Mischel, W. (2014). *The Marshmallow Test: Why Self-Control Is the Engine of Success*. New York: Little Brown.

Mitroff, I. I. (1972). The myth of objectivity, or why science needs a new psychology of science, *Management Science*, 18, B613–B618.

Mitroff, I. I., & Mason, R. O. (1974). On evaluating the scientific contribution of the Apollo Moon Missions via information theory: A study of the scientist-scientist relationship. *Management Science*, 20 (12), 1501–1513.

Mixon, D. (1972). Instead of deception. *Journal for The Theory of Social Behaviour*, 2 (2), 145–178.

Moher, D., Hopewell, S., Schulz, K. F., et al. (2010). CONSORT 2010 explanation and elaboration: Updated guidelines for reporting parallel group randomized trials. *British Medical Journal*, 340:c869. DOI: 10.1136/bmj.c869.

Munafo, M. R., Nosek, B.A., Bishop, D. V., et al. (2017). A manifesto for reproducible science. *Nature Human Behavior: Perspective*, 1 (0021), January 10.

Murray, C. (2016). *By the People: Rebuilding Liberty without Permission*. New York: Crown Forum.

Naftulin, D. H., Ware, J. E., Jr., & Donnelly, F. A. (1973). The Doctor Fox Lecture: A paradigm of educational seduction. *Journal of Medical Education*, 48 (7), 630–635.

National Health and Medical Research Council, the Australian Research Council and Universities Australia, & Commonwealth of Australia (2018). *National Statement on Ethical Conduct in Human Research 2007 (Updated 2018)*. Canberra, ACT: National Health and Medical Research Council.

National Research Council (2002). *Access to Research Data in the 21st Century: An Ongoing Dialogue among Interested Parties: Report of a Workshop.* Washington, DC: The National Academies Press. Available at https://doi .org/10.17226/10302

National Science Foundation (1953). *The Third Annual Report of the National Science Foundation: Year Ending June 30, 1953.* Washington, DC: US Government Printing Office.

National Science Foundation (2019). *Proposal and Award Policies and Procedures Guide.* February 25, NSF 19-1, OMB Control Number 3145-0058. Available at https://www.nsf.gov/publications/pub_summ.jsp?ods_key=papp

Nelson, R. R. (1959). The simple economics of basic scientific research. *Journal of Political Economy,* 67(3), 297–306. Available at www.jstor.org/stable/ 1827448

Newton, I. (1675). Letter from Sir Isaac Newton to Robert Hooke. February 5. Available at https://digitallibrary.hsp.org/index.php/Detail/objects/9792, January 23, 2019.

Newton, I. (1726). *Philosophiae naturalis principia mathematica,* 3rd ed. London: W. & J. Innys, Royal Society.

Newton, I. (1729). *The Mathematical Principles of Natural Philosophy,* Translated from 3rd ed. into English by Andrew Motte. London: Benjamin Motte.

Nijhawan, L. P., Janodia, M. D., Muddukrishna, B. S., et al. (2013). Informed consent: issues and challenges. *Journal of Advanced Pharmaceutical Technology & Research,* 4(3), 134–140.

Nobel Media AB (2005). *Nobel Prize in Physiology of Medicine* 2005: B. J. Marshall and J. R. Warren. NobelPrize.org, 3 October. Available at https:// www.nobelprize.org/prizes/medicine/2005/press-release/

Nosek, B. A., Bar-Anan, Y. (2012). Scientific utopia: I. *Opening scientific communication.,* Psychological Inquiry, 23, 217–243.

Nosek, B. A., Spies, J. R., & Motyl, M. (2012). Scientific utopia: II. *Perspectives on Psychological Science,* 7, 615–631.

O'Connor, J. J., & Robertson, E. F. (2004). The Royal Society. *MacTutor History of Mathematics Archive,* February. Available at https://mathshistory .st-andrews.ac.uk/Societies/RS/.

OECD (2003). *The Sources of Economic Growth in OECD Countries.* Paris: OECD Publications Service.

OED Online (2018). "scientific method, n." *Oxford University Press,* July 2018. Available at www.oed.com/view/Entry/383323.

Oehler, J. H. (1976). Experimental studies in Precambrian paleontology: Structural and chemical changes in blue-green algae during simulated fossilization in synthetic chert. *GSA Bulletin,* 87 (1), 117–129.

Office of Management and Budget (2018). *Office of Management and Budget Historical Tables*. Available at www.whitehouse.gov/omb/historical-tables/.

Ofir, C., & Simonson, I. (2001). In search of negative customer feedback: The effect of expecting to evaluate on satisfaction evaluations. *Journal of Marketing Research*, *38*, 170–182.

O'Keefe, D. J. (1999). How to handle opposing arguments in persuasive messages: A meta-analytic review of one-sided and two-sided messages. *Annals of the International Communication Association*, *22*(1), 209–249.

O'Keefe, D. J., & Jensen, J. D. (2006). The advantages of compliance or the disadvantages of noncompliance? A meta-analytic review of the persuasive effectiveness of gain-framed and loss-framed messages. *Communication Yearbook*, *30* (1), 1–43.

O'Leary, C. J., Willis, F. N., & Tomich, E. (1970). Conformity under deceptive and non-deceptive techniques. *The Sociological Quarterly*, *11 (1)*, 87–93.

Open Science Collaboration (2015). Estimating the reproducibility of psychological science. *Science*, *349* (6251), 1–8.

Oppezzo, M., & Schwartz, D. L. (2014). Give your ideas some legs: The positive effect of walking on creative thinking. *Journal of Experimental Psychology: Learning Memory, and Cognition*, *40* (4), 1142–1152.

ORSA Committee on Professional Standards (1971). Guidelines for the practice of operations research. *Operations Research*, *19*, 1123–1258.

Ortmann, A., & Gigerenzer, G. (1997). Reasoning in economics and psychology: Why social context matters. *Journal of Institutional and Theoretical Economics*, *53* (4), 700–710.

Orwell, G. (1945). Notes on nationalism. *Polemic, No 1*. Available at http://gutenberg.net.au/ebooks03/0300011h.html#part30.

Ostro, S. J. (1993). Planetary radar astronomy. *Reviews of Modern Physics*, *65* (4), 1235–1279.

Ostrom, E. (1990). *Governing the Commons: The Evolution of Institutions for Collective Action*. Cambridge, UK: Cambridge University Press.

Patterson, S. C., & Smithy, S. K. (1990). Monitoring scholarly journal publication in political science: The role of the *APSR*. *PS: Political Science & Politics*, 647–656.

Peirce, C. S. (1958). CP 7: Science and Philosophy. In A. W. Burks, ed., *Collected Papers*, EPUB, Cambridge, MA: Harvard.

Peters, D. P., & Ceci, S. J. (1982). Peer-review practices of psychological journals: The fate of published articles, submitted again. *Behavioral and Brain Sciences*, *5* (2), 187–195.

Platt, J. R. (1964). Strong inference. *Science*, *146*, 347–353.

Plavén-Sigray, P., Matheson, G. J., Schiffer, B. C., et al. (2017). The readability of scientific texts is decreasing over time. *eLIFE*, 27725. DOI: https://doi.org/10.7554/eLife.27725

PLoS ONE (2016a). *Submission Guidelines.* Available at http://journals.plos.org/plosone/s/submission-guidelines#loc- style-and-format.

PLoS ONE (2016b). *Criteria for publication.* Available at http://journals.plos.org/plosone/s/criteria-for-publication

Popper, K. (1959). *The Logic of Scientific Discovery.* London: Routledge.

Porter, M. (1980). *Competitive Strategy: Techniques for Analyzing Industries and Competitors.* New York, NY: Free Press.

Post, F. (1994). Creativity and psychopathology: A study of 291 world famous men. *British Journal of Psychiatry, 165* (1), 22–34,

Prasad, V., Vandross, A., Toomey, C., et al. (2013). A decade of reversal: An analysis of 146 contradicted medical practices. *Mayo Clinic Proceedings,* 88, 790–798. http://dx.doi.org/10.1016/j.mayocp.2013.05.012

Rasmussen, D. (2017). The gospel according to Michael Porter. *Institutional Investor,* November 8. Available at www.institutionalinvestor.com/article/b15jm11km848qm/the-gospel-according-to-michael-porter

Ravetz, J. (2004). The post-normal science of precaution. *Futures, 36,* 347–357.

Reid, L. N., Rotfeld, H. J., & Wimmer, R. D. (1982). How researchers respond to replication attempts. *Journal of Consumer Research,* 9 (2), 216–218.

Rich, B. R., & Janos, L. (1996). *Skunk Works: A Personal Memoir of My Years at Lockheed.* New York, NY: Little Brown and Company.

Ridd v. James Cook University (2019). BRG 1148 of 2017.

Ridley, M. (2020). *How Innovation Works.* London: 4th Estate.

Ring, K., Wallston, K., & Corey, M. (1970). Mode of debriefing as a factor affecting subjective reaction to a Milgram-type obedience experiment: An ethical inquiry. *Representative Research in Social Psychology, 1* (1), 67–88.

Routh, C. H. F. (1849). On the causes of the endemic puerperal fever of Vienna. *Medico-Chirurgical Transactions, 32,* 27–40.

Royal Society (2019). History of the Royal Society. *The Royal Society internet site.* Available at https://royalsociety.org/about-us/history/

Rushton, J. P., Murray, H. P., & Paunonen, S. V. (1987). Personality characteristics associated high research productivity. In D. N. Jackson & J. P. Rushton, eds., *Scientific Excellence: Origins and Assessment.* Thousand Oaks, CA: Sage, pp. 129–148.

Schachter, S. (1951). Deviation, rejection, and communication. *Journal of Abnormal and Social Psychology,* 46 (2), 190–207.

Scheibehenne, B., Greifeneder, R., & Todd, P. M. (2010). Can there ever be too many options? A meta-analytic review of choice overload. *Journal of Consumer Research, 37,* 409–425.

Schluter, D. (1994). Experimental evidence that competition promotes divergence in adaptive radiation. *Science,* 266 (5186), 798–801.

Schmidt, F. L. (1996). Statistical significance testing and cumulative knowledge in psychology: Implications for training of researchers. *Psychological Methods, 1* (2), 115–129.

Schmidt, F. L. (2017). Beyond questionable research methods: The role of omitted relevant research in the credibility of research. *Archives of Scientific Psychology, 5* (1), 32–41.

Schmidt, F. L. (2018). A theory of European Anti-Semitism. Downloaded January 30, 2019, from ResearchGate. Available at www.researchgate.net/publication/329487711_A_Theory_of_European_Anti-Semitism

Schmidt, F. L., & Hunter, J. E. (2015). *Methods of Meta-Analysis: Correcting Error and Bias in Research Findings*, 3rd ed. Thousand Oaks, CA: Sage.

Schmidt, F. L., Oh, I.-S., & Shaffer, J. A. (2016). The validity and utility of selection methods in personnel psychology: practical and theoretical implications of 100 years of research findings. *ResearchGate Working Paper*, DOI: 10.13140/RG.2.2.18843.26400.

Schmidt, S. (2009). Shall we really do it again? The powerful concept of replication is neglected in the social sciences. *Review of General Psychology, 13* (2), 90–100.

Schneider, C. E. (2015). *The Censor's Hand: The Misregulation of Human Subject Research*. Cambridge, MA: The MIT Press.

Schrag, Z. M. (2010). *Ethical Imperialism: Institutional Review Boards and the Social Sciences. 1965–2009*. Baltimore, MD: Johns Hopkins University Press.

Schrag, Z. M. (2014). You Can't Ask That. *Washington Monthly, September/October*.

Schroter, S., Black, N., Evans, S., et al. (2008). What errors do peer reviewers detect, and does training improve their ability to detect them? *Journal of the Royal Society of Medicine, 101*, 507–514. DOI: 10.1258/jrsm.2008.080062.

Schulz, K. F., Altman, D. G., Moher, D., for the CONSORT Group (2010). CONSORT Statement: Updated guidelines for reporting parallel group randomized trials. *PLOS Medicine, 7* (3), e1000251.

Shadish, W. R., Cook, T. D., & Campbell, D. T. (2001). *Experimental and Quasi-Experimental Designs for Generalized Causal Inference*, 2nd ed. Boston, MA: Cengage.

Sherif, M., Harvey, O. J., White, B. J., et al. (1961). *Intergroup Conflict and Cooperation: The Robbers Cave Experiment*. Norman, OK: University of Oklahoma Book Exchange.

Shih, Y., Huang, R. H. & Chiang, H. (2012). Background music: Effects on attention performance. *Work, 42*, 573–578.

Shulman, S. (2008). *The Telephone Gambit: Chasing Alexander Graham Bell's Secret*. New York, NY: Norton.

Silvia, P. J., & Phillips, A. G. (2013). Self-awareness without awareness? Implicit self-focused attention and behavioral self-regulation. *Self and Identity*, *12*, 114–127.

Simkin, M. V., & Roychowdhury, V. P. (2005). Stochastic modeling of citation slips. *Scientometrics*, *62*, 367–384.

Simon, J. L. (1996). *The Ultimate Resource 2*. Princeton, NJ: Princeton University Press.

Simon, J. L. (2002). *A Life Against the Grain*. London: Transaction Publishers.

Simon, R. J., Bakanic, V., & McPhail, C. (1986). Who complains to journal editors and what happens? *Sociological Inquiry*, *56* (2), 259–271.

Simonsohn, U. (2014). Citing prospect theory. Available at http://datacolada .org/15

Simonson, I., & Kivetz, R. (2018). Bringing (contingent) loss aversion down to earth – A comment on Gal & Rucker's rejection of "losses loom larger than gains." *Journal of Consumer Psychology*, *28* (3), 517–522.

Slovic, P., & Fischhoff, B. (1977). On the psychology of experimental surprises. *Journal of Experimental Psychology: Human Perception and Performance*, *3* (4), 544–551.

Smith, V. L. (1962). An experimental study of competitive market behavior. *Journal of Political Economy*, *70* (2), 111–137.

Smith, V. L. (1964). Effect of market organization on competitive equilibrium. *Quarterly Journal of Economics*, *78* (2), 181–201.

Smith, V. L. (1965). Experimental auction markets and the Walrasian hypothesis. *Journal of Political Economy*, *73* (4), 387–393.

Smith, V. L. (1991). *Papers in Experimental Economics (Collected Works)*. Cambridge: Cambridge University Press.

Smith, V. L. (1991). Rational choice: The contrast between economics and psychology. *Journal of Political Economy*, *99* (4), 877–897.

Smith, V. L. (2002). Method in experiment: Rhetoric and reality. *Experimental Economics*, *5*, 91–110.

Smith, V. L. (2003). Constructivist and ecological rationality in economics. *The American Economic Review*, *93(3)*, 465–508.

Smith, V. L. (2005). Behavioral economics research and the foundations of economics. *The Journal of Socio-Economics*, *34*, 135–150.

Smith, V. L. (2013). Adam Smith: From propriety and *Sentiments* to property and *Wealth*. *Forum for Social Economics*, *42* (4), 283–297.

Smucker, M. R. (2008). Michael J. Mahoney (1946–2006). *American Psychologist*, *63* (1), 53–54.

Soyer, E., & Hogarth, R. M. (2012). The illusion of predictability: How regression statistics misled experts. *International Journal of Forecasting*, *28*, 695–711.

Spencer, R. W., Christy, J. R., & Braswell, W. D. (2017). UAH Version 6 global satellite temperature products: methodology and results. *Asia-Pacific Journal of Atmospheric Sciences*, 53 (1), 121–130.

Sponsel, A. (2002). Constructing a "Revolution in Science": The campaign to promote a favourable reception for the 1919 solar eclipse experiments. *The British Journal for the History of Science*, 35 (4), 439–467.

Sprat, T. (1722). *The History of the Royal Society of London, For The Improving of Natural Knowledge*, 3rd ed. Samuel Chapman: London, England. Available at https://books.google.com.au/books?hl=en&lr=&id=u5dJAAAAcAAJ&oi=fnd& pg=PA1&dq=Thomas+Sprat+%22History+of+the+Royal+Society+of+1667% 22&ots=anb5aA5GDm&sig=1gSVSsVqp6lnLvdImBrCe8T-Jes&redir_esc= y#v=onepage&q=Thomas%20Sprat%20%22History%20of%20the%20Royal %20Society%20of%201667%22&f=false

Starbuck, W. H. (2006). *The Production of Knowledge*. Oxford University Press, Oxford

Staw, B. M. (1976). Knee-deep in the big muddy: a study of escalating commitment to a chosen course of action. *Organizational Behavior and Human Performance*, 16, 27–44.

Stewart, G. L. (2006). A meta-analytic review of relationships between team design features and team performance. *Journal of Management*, 32 (1), 29–55.

Stewart, W. W., & Feder, N. (1987). The integrity of the scientific literature. *Nature*, 325 (January 15), 207–214.

Straumsheim, C. (2017). The shrinking mega-journal. *Inside Higher Ed*, January 5, available at www.insidehighered.com/news/2017/01/05/open-access-mega-journal-plos-one-continues-shrink.

Strong, P. E. (2017). Wargaming the Atlantic War: Captain Gilbert Roberts and the Wrens of the Western Approaches Tactical Unit. *Validity and Utility of Wargaming, December 10th, 2017*, Paper for MORS Wargaming Special Meeting October 2017 – Working Group 2.

Styer, P., McMillan, N., Gao, F., et al. (1995). Effect of outdoor airborne particulate matter on daily death counts. *Environmental Health Perspectives*, 103 (5), 490–497.

Survey Research Center (1976). Research involving human subjects, 2 October 1976, box 11, meeting #23, tab 3(a). *National Commission for the Protection of Human Subjects of Biomedical and Behavioral Research, archival collection*, Bioethics Research Library Joseph and Rose Kennedy Institute of Ethics. Georgetown University, Washington, DC.

Sutton, R. I., & Rafaeli, A. (1988). Untangling the relationship between displayed emotions and organizational sales: The case of convenience stores. *Academy of Management Journal, 31* (3), 461–487.

Sveikauskas, L. (2007). R&D and productivity growth: a review of the literature. *BLS Working Papers*, Working Paper 408.

Szilard, L. (1961). *The Voice of the Dolphins and Other Short Stories.* New York, NY: Simon and Shuster.

Tangney, J. P., Baumeister, R. F., & Boone, A. L. (2004). High self-control predicts good adjustment, less pathology, better grades, and interpersonal success. *Journal of Personality*, 72 (2), 271–324.

Tetlock, P. E. (2005). *Expert Political Judgment?* Princeton, NJ: Princeton University Press.

Thornton, S. (2018). Karl Popper. *The Stanford Encyclopedia of Philosophy* (Fall 2018 ed.), Edward N. Zalta, ed. Available at https://plato.stanford.edu/

Trotter, W. (1941). The Collected Papers of Wilfred Trotter, F. R. S. London: Oxford University Press. Available at https://wellcomecollection.org/works/znsqw543/items?canvas=8

Tschoegl, A. E. & Armstrong, J. S. (2007). Review of Philip E. Tetlock, Expert political judgment: How good is it? How can we know? *International Journal of Forecasting*, 23 (2), 339–342.

Tufte, E. R. (2001). *The Visual Display of Quantitative Information.* Cheshire, CT: Graphics Press.

Tyler, J. E. (1834). *Oaths; Their Origin, Nature, and History.* London: J. W. Parker.

Warren, E. S. (2007). Michael J. Mahoney (1946-2006): A life celebration. *The Humanistic Psychologist*, 35 (1), 105–107.

Wason, P. C. (1960). On the failure to eliminate hypotheses in a conceptual task. *Quarterly Journal of Experimental Psychology*, 12 (3), 129–140.

Wasserstein, R. L., & Lazar, N. A. (2016). The ASA statement on *p*-values: context, process, and purpose. *The American Statistician*, 70 (2), 129–133.

Weisberg, D. S., Keil, F. C., Goodstein, J., et al. (2008). The seductive allure of neuroscience explanations. *Journal of Cognitive Neuroscience*, 20 (3), 470–477.

Weisberg, D. S., Taylor, J. C. V., & Hopkins, E. J. (2015). Deconstructing the seductive allure of neuroscience explanations, *Judgment and Decision Making*, 10 (5), 429–441.

Went, F. W. (1949). The plants of Krakatoa. *Scientific American*, 181 (3), 52–55.

Westfall, R. (1973). Newton and the fudge factor. *Science*, 179 (4075), 751–758.

White, B. L. (2002). *Classical Socratic Logic Provides the Foundation for the Scientific Search for Truth.* Oakland, CA: *Strategic Technology Institute*:

Wicherts, J. M., Borsboom, D., Kats, J., et al. (2006). The poor availability of psychological research data for reanalysis. *American Psychologist*, 61 (7), 726–728.

Williams, L. P. (2019). Michael Faraday. *Encyclopaedia Britannica*, February 21. Available at www.britannica.com/biography/Michael-Faraday.

Wilson, D. K., Purdon, S. E., & Wallston, K. A. (1988). Compliance to health recommendations: A theoretical overview of message framing, *Health Education Research*, 3 (2), 161–171.

Wilson, E. J., & Sherrell, D. L. (1993). Source effects in communication and persuasion: A meta-analysis of effect size. *Journal of the Academy of Marketing Science*, 21 (2), 101–112.

Wilson, J. D. (1978). Peer review and publication: Presidential Address before the 70th Annual Meeting of the American Society for Clinical Investigation, San Francisco, California, 30 April 1978. *Journal of Clinical Investigation, 61,* 1697–1701.

Winston, C. (2007), *Government Failure versus Market Failure.* Washington, DC, AEI-Brookings Joint Center for Regulatory Studies.

Wolf, A. (2002). *Does Education Matter? Myths about Education and Economic Growth.* London: Penguin.

Wright, M., & Armstrong, J. S. (2008). Verification of citations: Faulty towers of knowledge? *Interfaces, 38,* 125–139.

Wu, L., Wang, D., & Evans, J. A. (2019). Large teams develop and small teams disrupt science and technology. *Nature, 566,* Feb. 13, 378–382.

Wuchty, S., Jones, B. F., & Uzzi, B. (2007). The increasing dominance of teams in production of knowledge. *Nature,* 316 (5827), 18 May, 1036–1039.

Yankauer, A. (1985). Peering at peer review. *CBE Views,* 8(2), 7–10.

Yankauer, A. (1990). Who are the peer reviewers and how much do they review? *JAMA, 263* (10), 1338–1340.

Young, N. S., Ioannidis, J. P. A., & Al-Ubaydli, O. (2008). Why current publication practices may distort science. *PLOS Medicine, 5* (10), e201. doi:10.1371/journal.pmed.0050201.

Young, S. S., & Karr, A. (2011). Deming, data and observational studies: a process out of control and needing fixing. *Significance,* September, 116–120.

Zhou, J. (2003). When the presence of creative coworkers is related to creativity: Role of supervisor, close monitoring, developmental feedback, and creative personality. *Journal of Applied Psychology, 88* (3), 413–422.

Ziliak, S. T. (2011). Matrixx v. Siracusano and Student v. Fisher, Statistical significance on trial. *Significance,* 8 (3), 131–134.

Ziliak, S. T., & McCloskey, D. N. (2008). *The Cult of Statistical Significance: How the Standard Error Costs Us Jobs, Justice, and Lives.* Ann Arbor, MI: University of Michigan.

Zimmer, R. J., & Isaacs, E. D. (2014). Report of the Committee on Freedom of Expression. *uchicago.edu.* Committee on Freedom of Expression at the University of Chicago. Unpublished manuscript. Available at https://provost.uchicago.edu/sites/default/files/documents/reports/FOECommitteeReport.pdf

INDEX